2015
万达商业规划
销 售 类 物 业

WANDA COMMERCIAL PLANNING 2015
PROPERTIES FOR SALE

万达商业地产设计中心 主编

中国建筑工业出版社

EDITORIAL BOARD MEMBERS
编委会成员

主编单位
万达商业地产设计中心

规划总指导
王健林

执行编委
赖建燕 曲晓东 于修阳 吕正韬 黄国斌 尹强 林树郁
门瑞冰 张东光 王福魁 毛晓虎 杨旭 曾静 王群华

参编人员
叶啸 武春雨 俞小华 胡延峰 李子强 龚芳 孙志超
黄建好 杜文天 董华维 车心达 李暄荣 石亮 昌燕
赵宁宁 杨磊 栾赫 潘鸿岭 刘大伟 任睿 彭亚飞
刘晓敏 高维 周升森 薛瑜 陈晖 戚士林 李春阳
洪剑 漆国强 周鹏 马申申 范志满 孙一琳 张悦
杨健珊 朱镇北 梅帆 李鹏 桑国安 刘敏 王雅斯
纪文青 顾东方 陈海燕 张爱珍 王嵘 谭喜峰 刘征
薛勇 文善平 宋波 冯俊 霍雪影 邵强 张克 荣万斗
安云泽 李万顺 梁国涛

校对
陈文娜 周昳晗 刘保亮 申亚男 袁文卿 董根泉

CHIEF EDITORIAL UNIT
Wanda Commercial Estate Design Center

GENERAL PLANNING DIRECTOR
Wang Jianlin

EXECUTIVE EDITORIAL BOARD MEMBERS
Lai Jianyan, Qu Xiaodong, Yu Xiuyang, Lv Zhengtao, Huang Guobin, Yin Qiang, Lin Shuyu, Men Ruibing, Zhang Dongguang, Wang Fukui, Mao Xiaohu, Yang Xu, Zeng Jing, Wang Qunhua

PARTICIPANTS
Ye Xiao, Wu Chunyu, Yu Xiaohua, Hu Yanfeng, Li Ziqiang, Gong Fang, Sun Zhichao, Huang Jianhao, Du Wentian, Dong Huawei, Che Xinda, Li Xuanrong, Shi Liang, Chang Yan, Zhao Ningning, Yang Lei, Luan He, Pan Hongling, Liu Dawei, Ren Rui, Peng Yafei, Liu Xiaomin, Gao Wei, Zhou Shengsen, Xue Yu, Chen Hui, Qi Shilin, Li Chunyang, Hong Jian, Qi Guoqiang, Zhou Peng, Ma Shenshen, Fan Zhiman, Sun Yiling, Zhang Yue, Yang Jianshan, Zhu Zhenbei, Mei Fan, Li Peng, Sang Guoan, Liu Min, Wang Yasi, Ji Wenqing, Gu Dongfang, Chen Haiyan, Zhang Aizhen, Wang Rong, Tan Xifeng, Liu Zheng, Xue Yong, Wen Shanping, Song Bo, Feng Jun, Huo Xueying, Shao Qiang, Zhang Ke, Rong Wandou, An Yunze, Li Wanshun, Liang Guotao

PROOFREADERS
Chen Wenna, Zhou Yihan, Liu Baoliang, Shen Ya'nan, Yuan Wenqing, Dong Genquan

CONTENTS
目录

A FOREWORD 010
序言

EVOLUTION AND INNOVATION OF WANDA'S PLANNING DESIGN MANAGEMENT MODEL 012
万达规划设计管理的沿革与创新

B COMMERCIAL PROJECTS 014
商业项目

APPLICATION OF DMD METHOD ON THE MANAGEMENT OF PLANNING AND DESIGN 016
DMD 方法在规划设计管理中的应用

OUTSTANDING PROJECT
优秀项目

01 QUANZHOU PUXI WANDA MANSION 022
泉州浦西万达公馆

02 DONGGUAN HOUJIE WANDA LAKEVIEW MANSION 028
东莞厚街万达御湖公馆

03 NANNING QINGXIU WANDA MANSION 036
南宁青秀万达公馆

04 JIAMUSI WANDA PALACE 044
佳木斯万达华府

05 ANHUI BOZHOU WANDA PALACE 050
安徽亳州万达华府

06 JINSHAN WANDA PALACE 056
金山万达华府

07 YANTAI ZHIFU WANDA MANSION 062
烟台芝罘万达公馆

SALES OFFICE
售楼处

01 SALES OFFICE OF CHONGQING YONGCHUAN WANDA PLAZA 068
重庆永川万达广场售楼处

02 SALES OFFICE OF SICHUAN QINGYANG WANDA PLAZA 074
四川青羊万达广场售楼处

03 SALES OFFICE OF FUJIAN SANMING WANDA PLAZA 080
福建三明万达广场售楼处

04 SALES OFFICE OF BEIJING FENGKE WANDA PLAZA 086
北京丰科万达广场售楼处

05 SALES OFFICE OF CHANGDE WANDA LAKE MANSION 092
常德万达·湖公馆售楼处

06 SHANGHAI QINGPU WANDA MALL EXHIBITION CENTER 098
上海青浦万达茂展示中心

PROTOTYPE ROOM
样板间

01 PROTOTYPE ROOM OF NANPING WANDA CORE MANSION 104
南平万达中央华城样板间

02 PROTOTYPE ROOM OF NANNING JIANGNAN WANDA PLAZA 108
南宁江南万达广场样板间

03 PROTOTYPE ROOM OF FOSHAN SANSHUI WANDA PLAZA 112
佛山三水万达广场样板间

04 PROTOTYPE ROOM OF FUJIAN SANMING WANDA PLAZA 116
福建三明万达广场样板间

05 PROTOTYPE ROOM OF JILIN CHANGYI WANDA PLAZA 120
吉林昌邑万达广场样板间

06 PROTOTYPE ROOM OF YINCHUAN XIXIA WANDA PLAZA 124
银川西夏万达广场样板间

07 PROTOTYPE ROOM OF JIXI WANDA PALACE 128
鸡西万达华府样板间

08 PROTOTYPE ROOM OF TAIYUAN WANDA PLAZA 132
太原万达广场样板间

09 PROTOTYPE ROOM OF SHANGYU WANDA SOHO 136
上虞万达SOHO样板间

项目统计（新增项目）									
分类	2016年新项目统计						工时		
	项目总数	分区	项目数量	项目名称	交地月份	等效项目数	建筑	内装	景观	合计
		北	7	呼市玉泉★	11	0.09	52	39	36	127
				北京顺义★	12	0.05	26	20	18	64
				天津塘沽★	1	0.55	314	235	215	764
				西安高新★9-3	1	0.55	314	235	215	764
				西安高新9-5	1	0.55	217	130	152	499
				西安高新6	8	0.23	90	54	63	208
	11			合计			1041	732	717	2491

（表3）开发项目多地块与首开地块统计表

0.75。考虑到延续项目没有前期工作，原则上统一取0.33的折算系数。对于难点项目或调整较大的项目，个别分析并经设计中心总经理批准后可调高系数，原则上此系数不大于0.5。如：某项目11月份开业，其本年度的等效项目数为11/16=0.69，但在计算项目工时的时候需要多乘一个折算系数（表3）。

对于销售物业存在首开地块和后续地块工作量的差异，首开地块需要考虑售楼处、样板间甚至前期、方案的大量工作，但是后续地块前期工作相对较少，售楼处的工作量也不存在。为了准确统计工时，上表"★"项目定为新项目，未标注的为后续地块，统计工时的时候，根据工作量的不同，区别对待。

2. 标准工时

标准工时的确定，在分类统计时，又从时间维度、专业维度、业态维度上进行细化，如标准工时分类（图1）。

这些表格看似简单，但其填报整理的过程却必须经历反复修正，只有经过时间、具体项目案例多次检验的工时，才能最终形成真实有效的标准工时。为了校核标准工时的准确性，所有的工作事项都和《万达商业地产技术管控手册》的具体管控流程相挂钩，确保每一项工作都有明确的目的和意义（表4）。

3. 建立模型

将各种不同类型的标准工时分别乘以其对应项目的等效项目数，然后求和汇总就可以得到所需的总工时了。比如统计年度总工时（式1和表5），将所有类型都汇总就可以得到年度总工时了。当然也可以分区域、分阶段分别统计，表5中已经将不同阶段和不同区域的工时都分别提供可汇总统计的结果。

有了工时统计的结果，对应的人员需求就可以得出了（式2）。这样就建立了一个相对简单、行之有效的工时管理模型和人员需求模型。

年度总工时=∑（标准工时×等效项目数）　（式1）

需求人数=年度总工时÷人均年度标准工时　（式2）

在实际汇总时，还将非生产部门的工时也纳入到了

时间维度	持有：前期、总图、方案、初设、施工图、实施、业态调整
	销售：前期、总图、销售卖场、方案、方案深化、施工图、实施
专业维度	建筑、内装、景观、结构、机电
业态维度	持有：标准直投、非标直投、标准开发、非标开发、合作净地、合作改造
	销售：开发销售、文旅销售

（图1）标准工时分类

determined as per the month of land hand-over. Projects of direct investment whose project periods are not yet determined shall consider a period of 16 months with year-around operation and equivalent to 12/16=0.75. Specific proportion can be worked out as per the month of land hand-over. Development projects and properties for sale shall consider a period of 22 months. The tables shall be regulated after the definite project period is settled. For example, if the land of a project would be handed over in September, its equivalent project number in this year shall be 12/16=0.75 (Table2).

The number of equivalent project for continuing projects shall be determined as per the month of opening or joining in partnership. For projects that can't be completed in this year, the year of its completion shall be filled in. The equivalent project number of continuing projects shall adopt the same standard as the new projects and that of projects span across years shall be uniformly considers as 0.75. Given that continuing project doesn't involve preliminary work, a convert coefficient of 0.33 shall be taken in principle. For projects with more difficulties or greater adjustment, the coefficient can be raised with specific analysis and the approval of the general manager of the Design Center. In principle, the coefficient shall be no larger than 0.5. For example, if a project opens business in November, its number of equivalent project of this year shall be 11/16=0.69. However, when calculating the man-hour of the project, the final man-hour shall be the result of the original number multiples a convert coefficient (Table3).

As for the difference on the workload of initially-developed plot and continuing plot of properties for sale, the initially-developed plot shall take sales office, sample room and even a mass of preliminary work and scheme work into consideration, however, the continuing plot has little preliminary work and does not have the workload of sales office. In order to gather accurate man-hour statistics, the projects marked with "★" in the above table shall be settled as new projects and those unmarked as continuing projects. The man-hour for these two types shall be counted as per different standards based on different workload.

2. STANDARD MAN-HOUR

After standard man-hours are settled, they are further divided as per time dimension, specialty and type of business for classified statistics, i.e. the standard man-hour classification (Fig.1).

These tables appear to be simple at first sight, but actually they are repeatedly revised in the filling and organizing process. To achieve real and effective standard man-hours, man-hours must be checked for several times by time and specific projects. In order to verify the accuracy of the standard man-hours, all the work items are linked together with the specific control and management procedures specified in the Technical Management and Control Manual of Wanda Commercial Estate to ensure that each work has a specific objective and significance (Table4).

3. MODELING

The sum of the standard man-hours of various types multiply their corresponding number of equivalent project shall add up to the total man-hours needed. For example, by adding up the annual man-hours of all types, we can get the total annual man-hour (see Formula 1 and Table5). Without doubt, we can also count respectively according to regions and phases. In the table 5, summarized statistics results of man-hours are provided for different phases and different regions.

With the results of man-hour statistics, the corresponding staff demand can be worked out (see Formula 2). Thus, a relatively easy and effective man-hour management mode and staff

工作阶段	主要工作内容	单项工作工时		管控手册流程编号	管理标准
		工作事项分解			
1、前期阶段	地质灾害调研	审阅相关资料（地勘、现场照片、评估报告等）		1-3.1-1.1	5-3-3.1
		初步确定拟采取的措施；给发展部和建筑专业提出意见建议等		1-3.1-3.3.4	5-4-1.1
		个别项目需要外出参加专家论证会或外出调研			
	参与规划启动会	会议资料及前期准备		1-3.1-3.1.1	《规划设计启动会纪要》
		参加设计启动会			
		审阅会议纪要、审批OA			
	小计				
2、总图阶段	参与设计单位招标	招标计划、入围单位审核、招标文件制定、答疑		1-3.1-3.1.2	固定流程
		技术标评审：各家设计单位技术标书，技术表评分表		1-3.1-3.1.7	
		设计总包项目定标文件审核报批			
	总图、单体户型、竖向、立面签批移交配合	项目设计内控计划、模块化上线计划审批		1-3.1-3.1.1	1-3.1
		参与审核地下室范围（含签字）		1-3.1-3.1.11~15	2-2-1.0
		参与审核地下室平面（含签字）			
		单体平面方案签批（含签字）		1-3.1-3.5.1~2	
		参与总图竖向（含会议）		1-3.1-3.3.2	2-2-2.0
		参与审核外立面（含签字）		1-3.1-3.2.4	2-2-8.2

（表4）标准工时与管控流程挂钩

生产部门										副总及以下
中区				南区			文旅			
景观	建筑	内装	景观	建筑	内装	景观	建筑	内装	景观	合计
35	1081	176	23	1553	263	26	334	503	695	6677
255	1289	0	166	1648	0	220	9170	0	466	15047
146	30	79	47	0	0	0	1139	1272	429	3480
1606	2178	486	1223	3483	775	1691	6211	1446	3491	26226
1757	677	717	1144	922	1005	1535	119	308	1838	12068
694	1179	208	447	1683	409	605	119	450	105	8244
2005	1977	2038	1416	2846	2735	1822	569	4267	3287	28982
0	444	0	0	585	0	0	0	0	0	1700
468	936	468	468	936	468	468	1248	624	624	8112
										0
6966	9790	4172	4935	13656	5655	6367	18908	8871	10934	110534
3.34	4.69	2.00	2.36	6.54	2.71	3.05	9.06	4.25	5.24	52.94
	合计	9.05		合计	12.30		合计	18.54		
3	5	2	2	7	3	3	9	4	5	53
	合计	9		合计	12		合计	19		

（表5）区分生产阶段的生产部门年度工时汇总表

汇总表中。整个工时统计，通过分业态统计、分阶段统计，其结果互相对比、校核，确保数据准确；通过对生产部门和非生产部门工时的区分，得以灵活使用基本数据、避免因局部数据的变化改变整体结果后而导致对其他未变数据的影响。最终形成设计中心重要管理工具——《工时手册》，为规划设计管理的科学决策提供重要的数据支持。

三、决策

1. 通过统计发现问题

《工时手册》的基础是数据。数据来自于每个人对自己工作的描述；数据间的对比就直接地反映了工作差异。每个人对自己工作进行工时统计的过程，就是不断地总结、发现问题的过程，从而有助于每个岗位都能够自觉地发现问题、防患于未然。对数据的审核与整理有助于促成设计中心整体工作效率的提升，在对比与相互校核的过程中，发现更优的工作方法。

demand mode has been established.

Total annual man-hour =Σ (standard man-hour × number of equivalent project)   (Formula 1)
Staff demand = total annual man-hour ÷ annual standard man-hour per capita   (Formula 2)

Actually, the man-hours of non-production departments are also included into the summary table. The whole man-hour counting has been carried out as per type of business and phase, the results of which are compared and verified to ensure the accuracy of the data; through the division of production department and non-production department, a flexible use of basic data has been achieved, thus avoiding the effect on other unchanged data caused by the change of partial data that alters the overall results. Finally, the *Man-hour Manual*, which serves as an important management tool for the Design Center, has been compiled, providing essential data support for the scientific decision making concerning the management of planning and design.

III. DECISION MAKING

1. RECOGNITION OF PROBLEMS VIA STATISTICS

The basis of the *Man-hour Manual* is data. The data are gathered from the description that everyone made for their

2. 实现合理资源分配

工时统计的每一次结果，从设计中心整体的结果看，在部门间总能出现一些人力与项目的不匹配。年中，项目短期内大幅增加时，超过正常工作量的25%时，就要启动人力增编需求。人力的需求计划、预算的调整都由项目增加引起，但两者都要通过工时的计算来准确体现数据的变化。精确的数据让模糊的资源分配过程简明而合理。

3. 动态调整实时反馈

项目发展、开业时间基本按照年度计划进行，人力资源的配置也以年初的项目发展、开业计划而定；但项目发展、开业计划，基本上是每月调整一次。任何一个职能部门都不可能每月调整一次人力资源的分配，这就出现了短期内项目和人力需求的不匹配。这种暂时的不匹配，就可以通过"项目管控外"的任务来平衡。"项目管控外"的任务有很多种，如设计竞标、概念规划、BIM研发、规划系统的制度修编等工作。灵活分配这些工作，基本能够实现工作负荷与人力需求的动态调整。

4. 绩效考核数据支持

工时统计理论上可以直接和绩效考核挂钩，但绩效考核的涉及面很广，非量化指标较多，实际执行中会有很多的困难。用工时统计的数据反映每个岗位的工作饱满度是个可以采用的指标。用工作饱满度除以费用，得出各岗位的费效比，能够作为相对客观的量化指标。

结语

归纳起来，DMD工作法可取之处在于：数据来自于实际工作中的分析提炼；根据工作需要来构造模型并在工作中反复检验、修正；决策的依据来自于理性、量化、科学的数据和模型。虽然我们利用DMD的工作方法在规划设计管理上取得了一些成果，但这只是一个开始，在今后的工作中，还可以利用不断积累的实际管理经验，不断修正和完善《工时手册》，为更多的管理决策提供更多的数据支持，不断提升科学、高效的管理方法。

work; the comparison between the data directly reflects the difference of work. The process of every one counting his/her work hours is a process of continuously discovering and summarizing problems, which is helpful for each position to detect problems consciously and nip them in the bud. The check and organization of data helps to promote the work efficiency of the whole Design Center. In the process of comparing and mutual checking, better working procedures can be found.

2. REALIZATION OF REASONABLE RESOURCES DISTRIBUTION

Viewing the Design Center as a whole, we can always detect some mismatches of man power and project between departments from the results of the man-hour counting. In the middle of a year, when the number of projects increases sharply in a short time and the workload exceeds normal workload by 25%, there is need for adding man power. Although the man power demand plan and adjustment of budget are caused by the increase of project, the change of their statistics both need to be reflected via man-hour counting. Accurate data makes the blurry resource allocation process simple and clear.

3. DYNAMIC REGULATION AND REAL-TIME FEEDBACK

The time for project development and business opening basically sticks to the annual plan and human resources are also distributed as per the project development and business opening plan made at the beginning of the year; the project development and business opening plan is basically adjusted every month, but it is impossible for any functional department to reallocate human resources every month, which results in the mismatch between project and manpower requirement in a short term. This type of temporary mismatch can be balanced through tasks that are not included in project management and control. There are a variety of tasks that are not included in project management and control, i.e., design bidding, conceptual plan, BIM research and development, system editing and revision of planning system. Dynamic regulation of workload and manpower requirements can be realized via a flexible allocation of these tasks.

4. DATA SUPPORT FOR PERFORMANCE ASSESSMENT

Theoretically speaking, man-hour counting can be directly linked together with performance assessment. However, since the coverage of performance assessment is quite extensive with various non-quantitative indexes, it is hard to actually put it into practice. The work fullness of each position reflected by the data of man hour counting is an adoptive index. Dividing the work fullness by cost, we can get the cost-benefit ratio of each position, which can be taken as a relatively objective quantitative index.

CONCLUSION

In conclusion, the merit of DMD method lies in that the data is derived from and analyzed based on actual work; the model is established based on work demand and is repeatedly verified and revised during the work; the basis for decision making is reasonable, quantified and scientific data and model. Although we have made some achievements on the management of planning and design with DMD method, this is still a beginning. In the future, with the constantly accumulated practical management experience, continuous revision and perfection shall be made for the Man-hour Manual to provide more statistical support of management decision making and constantly improve scientific and efficient management method.

优秀项目

01 QUANZHOU PUXI WANDA MANSION
泉州浦西万达公馆

时间：2015 / 09 / 30　　**OPENED ON**：30th SEPTEMBER, 2015
地点：福建 / 泉州　　　**LOCATION**：QUANZHOU, FUJIAN PROVINCE
占地面积：18.27公顷　　**LAND AREA**：18.27 HECTARES
建筑面积：60万平方米　　**FLOOR AREA**：600,000M²

01 泉州浦西万达公馆总平面图
02 泉州浦西万达公馆外立面
03 泉州浦西万达公馆入口

PROJECT OVERVIEW
项目概况

泉州浦西万达广场位于泉州市丰泽区中心区域沿江地段，地理位置优越。规划用地18.27公顷，总建筑面积123万平方米，是一个集五星级酒店、甲级写字楼、商务酒店、名牌商店、高档电影城、高级写字楼、高档住宅、高档百货、娱乐休闲、餐饮和生活配套等多功能为一体的"万达城市综合体"，已成为泉州城市中心地标性的建筑。万达公馆为泉州市第一个高水平精装豪宅，建筑面积60万平方米。

Ideally located in the central area of Fengze District, Quanzhou and the neighboring river, Quanzhou Puxi Wanda Plaza enjoys superior geographical location. With its planned site area of 18.27 hectares and gross floor area of 1,230,000 square meters, the Plaza, a landmark building in Quanzhou city center, is served as Wanda Urban Complex that integrates multiple functions into one, such as a five-star hotel, a grade-A office building, a business hotel, brand stores, a high-end cinema, a high-end office building, a high-end residence, a high-end department store, recreation & entertainment, catering and living facilities. Wanda Mansion, with gross floor area of 600,000 square meters, is the first high-end classic luxury in Quanzhou City.

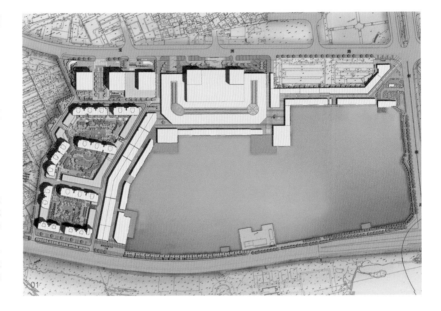

04

ARCHITECTURAL PLANNING
建筑规划

建筑整体规划采用U型布局结构，最大化利用江景、湖景；半围合大庭院提供舒适的公共空间和室外风景，既能充分享受到自然阳光、又可内院观赏湖景。建筑采用ArtDeco建筑风格，强调竖向线条挺拔向上，寓意不断超越的人文精神。整体化的风格，使得高密度建筑群对城市肌理的侵扰被弱化，仿佛是江边一块纯洁的幕布，对泉州城市环境提升贡献了很大力量。

The general architectural planning in U-shaped layout maximizes the use of river view and lake view, and applies semi-enclosed courtyard to both render comfortable public space & outdoor scenes and enjoy sunshine and lake view at inner courtyard. ArtDeco style highlights tall and straight vertical lines, implying unceasingly surmounted humanistic spirit. Integration planning weakens intrusion of high-density buildings to city texture and makes them an unstained curtain erecting along the river, which contribute a lot to enhancing the environment of Quanzhou City.

05

04 泉州浦西万达公馆外立面
05 泉州浦西万达公馆夜景
06 泉州浦西万达公馆景观方案设计图
07 泉州浦西万达公馆中央水景

06

LANDSCAPE DESIGN
景观

景观设计将"欧式风情"进行元素提炼，以"阳光"、"自然"、"艺术"为关键词，精心打造各具特点的欧式庭园。空间通过对称的轴线关系，用暖色调的材质，营造出既大气磅礴又温馨怡人的氛围。园林绿化模拟自然生态进行布置，讲求乔、灌、草的科学搭配，创造"春花、夏荫、秋实、冬青"的景观效果。整体环境与富有意境的景观节点相融合，给人们提供一个方便、舒适、优美的居住场所，于是住宅区便是一道亮丽的风景。

To well construct distinctive European style gardens, the landscape design draws on European style elements and sticks to the key words of "sunshine", "nature", and "art". Application of symmetrical axis and warm-color materials is helpful to create majestic yet cozy spatial atmosphere. Landscaping laid out by simulating nature pursues scientific collocation of trees, shrubs and grasses, and achieves landscape effect featuring spring blossom, summer shade, autumn fruit and winter green. When the whole environment is melted into landscape nodes, the convenient, comfortable and graceful residential area becomes a splendid landscape.

INTERIOR DESIGN
内装

室内精装定位为新古典主义风格。淡雅的卡布其诺色调，搭配"点睛"的金色配饰，在柔和的灯光下静静地释放出独特的浪漫气息。天然大理石的自然纹理、精雕细琢的装饰线条与华丽的手绘壁纸相呼应，搭配手工定制家具饰品，营造具有宫廷般"古典美"的空间。

The interior fitting-out pursues the neoclassical style. The simple and elegant cappuccino color going with focused gold accessories quietly creates a unique romantic atmosphere in the soft light. The harmony between natural texture and exquisite decorative lines of natural marble and magnificent hand-painted wallpaper, and the display of hand-tailored furniture accessories together build the court-like space of the classical beauty.

08

08 泉州浦西万达公馆大堂设计图
09~10 泉州浦西万达公馆客厅
11 泉州浦西万达公馆书房
12 泉州浦西万达公馆餐厅
13 泉州浦西万达公馆卧室

优秀项目

02 DONGGUAN HOUJIE WANDA LAKEVIEW MANSION
东莞厚街万达御湖公馆

时间：2015 / 03 / 28　　**OPENED ON**：28th MARCH, 2015
地点：广东 / 东莞　　　　**LOCATION**：DONGGUAN, GUANGDONG PROVINCE
占地面积：13.98公顷　　**LAND AREA**：13.98 HECTARES
建筑面积：24.95万平方米　**FLOOR AREA**：249,500M²

PROJECT OVERVIEW
项目概况

东莞厚街万达广场位于东莞市厚街镇莞太路与体育路交汇处，是万达集团继长安、东城万达广场之后，全面布局东莞的又一力作。东莞厚街万达广场定位为城市综合型商业中心，项目用地面积约13.98万平方米，总建筑面积达61.48万平方米，集15万平方米大型购物中心、滨湖商业街、高端住宅、精装SOHO等业态组成。其中，影城、电玩场、儿童乐园等娱乐业态达2万平方米，是厚街地区的"首席室内娱乐中心"；室内外餐饮街达1.5万平方米，打造厚街美食新中心。

Dongguan Houjie Wanda Plaza is located in the intersection between Guantai Road and Tiyu Road, Houjie Town, Dongguan City, and is another masterpiece of Wanda Group following the construction of Chang'an Wanda Plaza and Dongcheng Wanda Plaza of Wanda Group. With a site area of 139,800 square meters and gross floor area of 614,800 square meters, the Plaza is set to be an integrated commercial center of the city and incorporates a large-scale shopping center of 150,000 square meters, a high-end residence, a refined-decorated SOHO and other business contents. Among them, entertainment contents such as cinema, super player and kids place occupy 20,000 square meters, making the Plaza the "No.1 interior entertainment center" in Houjie. And the interior and exterior food streets covering 15,000 square meters become a new food center of Houjie.

01

02

01 东莞厚街万达御湖公馆总平面图
02-03 东莞厚街万达御湖公馆外立面

ARCHITECTURAL PLANNING
建筑规划

公馆方案立面采用纯正的ArtDeco建筑风格，着眼于现代风格演化的立面效果，使得轮廓分明，竖向分隔挺拔内敛。建筑整体造型端庄稳重，基座、墙身、檐口屋顶，色彩统一中又略有变化，使之稳重而不沉重，造就了居住区宁静祥和的气氛。

Authentic ArtDeco style facade of the Mansion pursues modern style effect, and results in distinct profile and towering & understated vertical division. The whole building looks dignified, with its base, wall, eave and roof unified yet slightly changeable in color to present such dignified yet not ponderous gesture and quiet & peaceful residential atmosphere.

04 东莞厚街万达御湖公馆夜景

PART B | COMMERCIAL PROJECTS 商业项目 | 031

LANDSCAPE DESIGN
景观

住宅区紧邻中央人工湖，与商业部分隔水相望。项目动静分离，既不失商业氛围，又同时兼顾住宅区的私密性，正所谓"闹中取静"之高端度假休闲居所。住宅的景观空间由"巴比伦空中花园"概念结合东南亚风情元素而成。围绕中央湖泊营造出具有度假酒店风格的屋顶花园及下沉台地式景观空间，为业主提供堪比酒店度假的生活品质。同时，本案景观设计既要打造出一个品质生活场所，又要拉开与其他项目的差异性。为此，本设计深度挖掘景观优势来营造出韵味十足的高端品质度假区的感受。

The residential area is next to the central artificial lake and faces the commercial area across the lake. With dynamic (commercial atmosphere)-static (residential privacy) separation, the Project is thus said to be a high-end tranquil leisure resort amid chaos. The landscape space of the residence combines the concept of the Hanging Gardens of Babylon and the Southeast Asia style elements. It encompasses the roof garden specific to resort hotel and sunken platform type landscape space around the central lake, providing clients with quality life resembling resort hotel. To make the Project distinctive, the landscape design goes deep into landscape superiority to deliver an atmosphere that is compared to a high-end quality resort of lasting appeal.

05 东莞厚街万达御湖公馆绿化
06 东莞厚街万达御湖公馆景观亭
07 东莞厚街万达御湖公馆花坛
08 东莞厚街万达御湖公馆景观
09 东莞厚街万达御湖公馆绿化

10

10 东莞厚街万达御湖公馆公共入户大堂
11 东莞厚街万达御湖公馆分户入户大堂
12-13 东莞厚街万达御湖公馆公共入户大堂

11

优秀项目

05 ANHUI BOZHOU WANDA PALACE
安徽亳州万达华府

时间：2015 / 12 / 15	**OPENED ON**：15th DECEMBER, 2015
地点：安徽 / 亳州	**LOCATION**：BOZHOU, ANHUI PROVINCE
占地面积：20.0 公顷	**LAND AREA**：20.0 HECTARES
建筑面积：90.0 万平方米	**FLOOR AREA**：900,000M²

PROJECT OVERVIEW
项目概况

安徽亳州万达广场位于安徽省亳州市，规划区域20.0公顷，总建筑面积约90.0万平方米，由商业综合体、乙级写字楼、商务公寓、室外步行街、特色商业街及住宅等业态组成；集文化艺术、旅游休闲、商业娱乐、商务办公和高品质居住"五大功能"于一体，亳州万达华府为亳州市南部片区高品质精品住宅。

Anhui Bozhou Wanda Plaza is located in Bozhou City, with a planned area of 20.0 hectares and gross floor area of 900,000 square meters. The Plaza consists of such commercial contents as a commercial complex, a grade-B office building, a business apartment building, an exterior pedestrian street, a featured commercial street and a residence, and integrates five functions (i.e. culture and art, tourism and leisure, business and entertainment, business office and high-end residence) into one. Bozhou Wanda Palace is a high-quality residence in southern district of the city.

01 安徽亳州万达华府外立面
02 安徽亳州万达华府入口
03 安徽亳州万达华府总平面图

ARCHITECTURAL PLANNING
建筑规划

整体规划布局分区明确，住宅以大花园为中心，四周布置沿街商业，有效地带动商业氛围和住宅配套。建筑以新古典建筑风格为主，底商及金街突出时代感和创新性。整个建筑群体高低错落，天际线变化有致，立面凹凸有律，变化中富有节奏。建筑采用"三段式"对称的立面构图原则。高层住宅主体以浅色涂料为饰面，住宅基座饰以石材。石材以暖调的褐色为基调，突出端庄、高雅的风范。建筑整体形态简洁自然而又精致典雅，成为亮丽的地标！

With a clearly defined planning layout, residence centers on garden and is surrounded by business along the street, which effectively mobilizes commercial atmosphere and residence auxiliary facilities. Buildings pursue the Neoclassical style and commerce at the bottom and golden street exhibits sense of the times and innovativeness. Well-arranged buildings, changeable skyline and uneven facade all seek rhythmical change. Adopting the "three-section" and symmetrical facade composition principle, main structure of high-rise residence is applied with light color coating and residence base is decorated with stone mainly in brown to deliver a dignified and elegant style. As a result, building in whole presents simple and natural yet delicate and elegant shape, making it a beautiful landmark.

LANDSCAPE DESIGN
景观

住宅区景观设计风格为欧式自然，是一种既具有庄重、有序列感的轴线处理，也具有自然和自由的庭院布置的风格。这种具有节奏感的处理手法由园区内婉转的流水贯穿其中，为劳累了一天回家的业主提供了从商业氛围中，逐渐拉回到休闲居家的安心体验；随处可见的自然点缀和高差不同的硬质平台相结合，再加上精心设置的休闲场地和娱乐场所，为老人及儿童提供了多功能的活动方式。

The landscape design pursues European natural style, which pays attention to both solemn & orderly axis treatment and natural & unconstrained courtyard layout. Winding water runs through the rhythmical style, relieving tired clients from commercial work. Natural ornaments scattered everywhere, rigid platforms of different height and well-arranged leisure and entertainment venues offer the old and children diverse activities.

04 安徽亳州万达华府外立面
05 安徽亳州万达华府绿化
06 安徽亳州万达华府水景

INTERIOR DESIGN
公馆内装

室内设计采用传统欧式的基调，融入现代元素的处理手法，既体现欧式风格的典雅，又展现当代豪宅舒适细腻的风范。设计表现上，欧式风格带来视觉上的绝对华丽，每一处细节都散发精湛工艺的居住魅力，方寸之间皆奢华感与艺术感。空间功能设计上，将传统的欧式尊贵性与现代的实用性完美地结合，带给高端业主独一无二的尊享感。

The interior design incorporates the modern elements into the traditional European style, which presents both elegance of the European style and comfort of modern mansions. In terms of design expression, the European style is visually magnificent, with every detail carrying exquisite residence appeal and being luxurious and artistic. In terms of the functional design, high-end owners may have unique experience, through ideal combination of the dignified European style and pragmatic modern style.

07 安徽亳州万达华府卧室
08 安徽亳州万达华府餐厅
09 安徽亳州万达华府客厅
10 安徽亳州万达华府卧室

06 JINSHAN WANDA PALACE
金山万达华府

时间：2015 / 07 / 17
地点：上海 / 金山区
占地面积：11.4 公顷
建筑面积：45.77 万平方米

OPENED ON : 17th JULY, 2015
LOCATION : JINSHAN DISTRICT, SHANGHAI
LAND AREA : 11.4 HECTARES
FLOOR AREA : 457,700M²

PROJECT OVERVIEW
项目概况

金山万达广场是万达集团在上海建造的第6个万达广场，位于上海市金山区杭州湾大道与龙皓路交叉口西南角，总占地面积11.4公顷，总建筑面积达45.77万平方米，由商业广场、精装SOHO和高档住宅三大部分组成；从功能上可划分为室内步行街、室外金街、精装SOHO和高档住宅，从体量来看将是全市万达广场中最大的一个。广场紧邻金山区政府，交通便利，周边的住宅区分布密集，使得该项目具有巨大的商业潜力，将成为整个区域的城市地标。

Being the sixth Wanda Plaza constructed by Wanda Group in Shanghai, Jinshan Wanda Plaza is located in southwest corner of intersection between Hangzhou Bay Avenue and Longhao Road, Jinshan District, Shanghai. With a site area of 11.4 hectares and gross floor area of 457,700 square meters, the Plaza comprises of three parts: a commercial plaza, a well-decorated SOHO and a high-end residence, and is functionally divided into an interior pedestrian street, an exterior golden street, a well-decorated SOHO and a high-end residence. Thus, the Plaza is the largest Wanda Plaza in Shanghai in terms of size. It boasts of convenient transportation as it is adjacent to Jinshan District Government and huge commercial potential as it is surrounded by densely distributed residential area. The Plaza is set to be a city landmark of the whole region.

01 金山万达华府总平面图
02 金山万达华府外立面
03 金山万达华府室外步行街

01
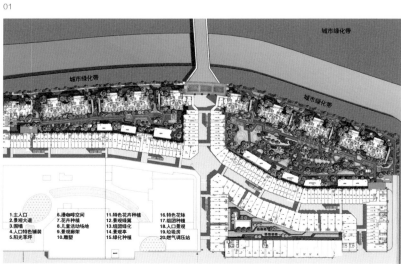

PART **B** | COMMERCIAL PROJECTS
商业项目

ARCHITECTURAL PLANNING
建筑规划

整体规划布局一字排开，南北通透，建筑沿天然河道南侧布置，不同高度的住宅有机结合，便捷的小区弧形路，形成了丰富的建筑空间。建筑风格采用国际流行的ArtDeco风格。风格简洁，强调垂直线条，灵活运用重复、对称、渐变等美学法则使几何造型充满诗意和富于装饰性。

Buildings are lined along the south side of the natural river course and enjoy light permeability in the south-north direction. Through an organic combination of residence of different height and convenient curve road, a rich space is achieved. The ArtDeco architectural style is applied to emphasize vertical lines and mould poetic and decorative geometric modeling by flexibly employing aesthetic principles of repetition, symmetry and gradation.

04 金山万达华府外立面
05 金山万达华府景观
06 金山万达华府花坛
07 金山万达华府雕塑小品

05

LANDSCAPE DESIGN
景观

景观设计呼应建筑立面从"丝绸"得来的造型灵感，经过对上海的文化和历史加以筛选和提炼，确定了"海上丝绸之路"的景观概念；然后提取浓缩为3个景观主题——"扬帆起航"、"乘风破浪"和"辉煌传奇"。每一个主题都有一个代表性的场景空间表达故事精神。广场景观呈现场景化，为参与和互动提供了条件，通过空间上有意识的围合与退让、流线引导、适当遮蔽等设计手法，唤起人们融入其中的激情。

Following the idea of the facade design and through filtering and refining culture and history of Shanghai, the landscape design settles its concept as a "Maritime Silk Road", and abstracts it into three themes, respectively being "setting sail", "riding on the wind" and "glorious legend". Each theme is provided with a representative scene space to deliver the spirit of the corresponding story. Besides, the landscape space practices scenario principle to pave the way for participation and interaction. Through conscious enclosure and recession, streamline guidance, appropriate concealment and other design methods, people are eager to get involved.

06

07

INTERIOR DESIGN
内装

样板间A户型设计风格为黄绿色调小清新美式风格,在硬装的配置上以浅色调为主,地面简洁的方形线条拼花形式、墙面白色背景线条的装饰,让现代美式风格更加凸显。在家具的选型上,注重线条更加的简练自然,布艺品质逸然,色调明亮清新,几抹暖黄的出现,让家萦绕平易近人的亲切感,透过直线条茶几玻璃镜面的层层折射,让周遭布艺颜色更加丰富,美式风情也更加清晰。

The Typical Plan A of the prototype room pursues the fresh American style in yellow green. In terms of hard decoration dominated by light color, floor simply paved in square lines pattern and wall decorated with white background lines highlight the modern American style. In terms of furniture, concise and natural lines, fabrics with elegant quality, and bright and fresh colors are preferred. The occasionally warm yellow color brings home intimacy. Reflected by glass of end table in straight lines, fabrics seem to be more colorful and the American style is more conspicuous.

08

08 金山万达华府餐厅
09 金山万达华府客厅
10 金山万达华府卧室
11 金山万达华府书房
12 金山万达华府客厅

09

金山样板间C户型整体设计风格为后现代简欧风格，硬装方面作为简欧派风格，整体色调以浅色为主，墙面注重简易线条装饰为主，同时融入欧式壁纸装饰，家具整体色系选择上注重跳色，适当增加明镜等软装挂件使整个空间感觉更为年轻有张力，轻盈明确，作为点缀色红色水晶灯的出现，跳耀着观者的视觉。简欧线条作为主题元素在空间内各个细节都得以彰显，将空间有机地整合在一起，并将奢华尺度渗透入每一个细节处。

The Typical Plan C of the prototype room pursues the postmodern Modern European Style. In terms of hard decoration dominated by light color, the walls are largely decorated by simple lines and the European style wallpapers. Furniture applies alternated colors and properly incorporates bright mirror, making the whole space more vigorous, stretching, graceful and distinctive. The red crystal lamp is visually attractive. The Jane European lines, the theme element, are found in every detail, organically integrating the whole space and infusing luxury scale.

优秀项目

07 YANTAI ZHIFU WANDA MANSION
烟台芝罘万达公馆

时间：2015 / 12 / 30　　**OPENED ON**：30th DECEMBER, 2015
地点：山东 / 烟台　　　**LOCATION**：YANTAI, SHANDONG PROVINCE
占地面积：4.61 公顷　　**LAND AREA**：4.61 HECTARES
建筑面积：15.06 万平方米　**FLOOR AREA**：150,600M²

PART B | COMMERCIAL PROJECTS 商业项目 | 073

INTERIOR DESIGN
售楼处内装

售楼处内装设计在欧式主调上同样渗入了地中海主题，门廊、穹顶等元素在内装设计中亦得到充分发挥。地面拼花、天花、石材墙、柱以及线条等细节精雕细琢，彰显了万达一贯的品质追求。立面通透的玻璃窗，配合大型的水晶吊灯，与墙面装饰形成一体的LED屏，使得售楼处室内空间既具有欧式的华贵，也具备良好的商务氛围。

While insisting on the European style in general, the interior design of the sales office adds the Mediterranean elements again. Porch, dome and other elements are frequently found in the design. Exquisite details such as parquet floor, ceiling, stone wall, column and mouldings show Wanda's consistent pursuit of quality. The transparent glass windows at the facade, together with the large crystal chandelier and LED screen integrated with wall decoration deliver both the European luxury and desirable commerce atmosphere.

07 重庆永川万达广场售楼处前厅接待台
08 重庆永川万达广场售楼处洽谈区
09 重庆永川万达广场售楼处大厅

02 SALES OFFICE OF SICHUAN QINGYANG WANDA PLAZA
四川青羊万达广场售楼处

售楼处

时间：2015 / 04 / 25　　**OPENED ON** : 25ᵗʰ APRIL, 2015
地点：四川 / 成都　　　**LOCATION** : CHENGDU, SICHUAN PROVINCE
建筑面积：899.5平方米　**FLOOR AREA** : 899.5M²

ARCHITECTURAL DESIGN
售楼处建筑

青羊万达广场售楼处布置于主建筑群南侧，邻近样板间；售楼处与样板房的交通组织形成"进入—集散—流动"的序列。建筑立面为现代简约构成主义风格，与背景主建筑群形成统一的城市界面形态。屋面及入口雨棚铝板线条构成了流畅的建筑轮廓，主立面采用三角体拼接面玻璃幕墙与平面玻璃幕墙交接，用构成主义的完美比例取得与整体轮廓相得益彰的效果。

The sales office of Qingyang Wanda Plaza is located in the south of the main buildings and adjacent to the prototype room. The sales office and prototype room present traffic organization order of an "entry-gathering-circulation". The building facade in the modern simple constructivism style and the background of main buildings constitute a unified city interface. Roof and entrance canopy aluminum plate lines depict smooth architectural profile. The main facade adopts triangle jointed glass curtain wall to connect with plane glass curtain wall, and interact with overall profile through ideal proportion of the constructivism.

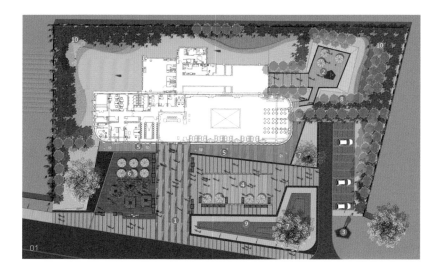

01 四川青羊万达广场售楼处总平面图
02 四川青羊万达广场售楼处外立面

LANDSCAPE DESIGN
售楼处景观

景观结合建筑，以"竹韵青羊，雅乐万达"为主题，力求使场地完整统一。根据需求设计了迎宾广场、休闲活动区、停车区及样板间景观区等功能区。铺地与草地边界交错、配合设置石小品演绎建筑的线条、色彩、元素等；同时还设置了景观镜面水景，使水景与建筑交相辉映。

In combination with buildings, the landscape follows the theme of "Bamboo in Qingyang, Music in Wanda" to achieve a complete and uniform site. The functional areas are designed as needed, such as an Overture Plaza, a recreation area, a parking area and a prototype room and a landscape area. Overlapped pavement and lawn boundary and stone landscape pieces together interpret lines, colors and elements of buildings. Besides, mirror waterscape is provided to eco with buildings.

03 四川青羊万达广场售楼处景观
04 四川青羊万达广场售楼处绿化
05 四川青羊万达广场售楼处景观
06 四川青羊万达广场售楼处水景

PART B | COMMERCIAL PROJECTS
商业项目

04

05

06

INTERIOR DESIGN
售楼处内装

售楼处室内虽为简欧风格,但突破传统简欧风格的繁琐,使整个空间更为硬朗、简洁。设计把成都当地元素融入其中,大面积的米色天然石材及仿石材墙砖,搭配"点睛"的白色GRG线条,在柔和的暖灯光下显现空间的简约而奢华。天然大理石的纹理、精雕细琢的装饰线条、华丽的花型水晶灯具,搭配欧式家具饰品,营造出兼具"东西方合璧之美"的室内空间。

Although pursuing the Modern European Style, the interior design gets rid of its conventional cumbersomeness to make the whole space more rigid and concise. The largely applied beige natural stones and stone like wall bricks going with focused white GRG lines creates a concise yet luxurious space under the soft light. The texture of natural marble, exquisite decorative mouldings, magnificent flower-shaped crystal lamps, and the European furniture accessories together build the space of blended Sino-Western beauties.

07 四川青羊万达广场售楼处沙盘
08 四川青羊万达广场售楼处前厅接待台

PART B COMMERCIAL PROJECTS
商业项目

091

10

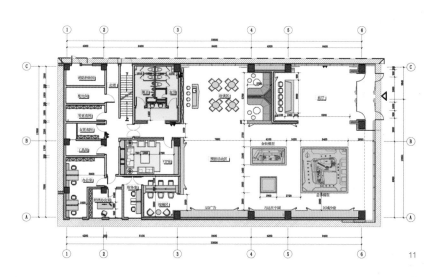

11

INTERIOR DESIGN
售楼处内装

内装设计通过"互联网+"的设计语言，将室内与室外做了良好的衔接。墙面上采用"互联网+"符号的夹胶玻璃，地面采用微晶砖拼接，使得整个空间科技感十足；顶棚水波纹柔美的造型也为展示中心增加了亲和力。

The interior design well links the interior and exterior via the design language of "Internet+". Laminated glasses with "Internet+" symbol on wall and microlite tile-paved floor contribute to a space full of technological aesthetics. The ceiling in graceful ripple shape makes the exhibition center more intimate.

09 北京丰科万达广场售楼处大厅
10 北京丰科万达广场售楼处前厅接待台
11 北京丰科万达广场售楼处平面图

售楼处

05 SALES OFFICE OF CHANGDE WANDA LAKE MANSION
常德万达·湖公馆售楼处

时间：2015 / 04 / 15　　**OPENED ON**：15th APRIL, 2015
地点：湖南 / 常德　　　　**LOCATION**：CHANGDE, HUNAN PROVINCE
建筑面积：4000 平方米　**FLOOR AREA**：4,000M²

01 常德万达·湖公馆售楼处外立面
02 常德万达·湖公馆售楼处设计手绘图
03 常德万达·湖公馆售楼处总平面图

ARCHITECTURAL DESIGN
售楼处建筑

售楼处采用简洁的ArtDeco新古典主义风格，在典雅的比例尺度基础上结合现代设计手法，融合民族风格的优点，追求时尚与动感设计，注重色彩表现，体现项目地标效果，同时与周边建筑保持和谐相处。立面强调建筑整体形体的变化和空间比例尺度的运用，弱化装饰构件；材质方面，玻璃幕、金属铝板和石材相互交融，将造型与功能完美相结合。在整体沿街效果上，通过多种模块的排列和组合形成丰富、自然的变化，对重点立面节点（如转角处的大玻璃面设计、售楼处入口、样板间入口等）特殊设计，强化了售楼处沿街立面效果。

Pursuing the simple ArtDeco neoclassical style, the sales office combines modern the design methods and merits of the national style based on elegant ratio scale, to seek chic and dynamic design, highlight color expression, display landmark effect and live in harmony with surrounding buildings. The facade attaches great importance to overall shape variation and spatial ratio scale, and weakens decorative components. Interaction of glass curtain, metal aluminum plate and stone ideally integrates modeling and function demands. Thanks to enriched and natural changes produced by permutations and combinations of the diverse modules, and specific design of the key facade nodes (e.g. the oversized glass at corner, entrance of sales office, entrance of the prototype room), frontage facade effect is intensified.

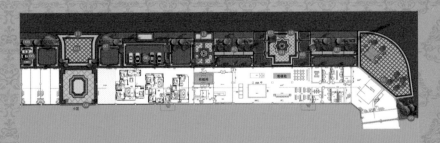

04

04 常德万达·湖公馆售楼处设计手绘图
05 常德万达·湖公馆售楼处外立面
06 常德万达·湖公馆售楼处景观
07 常德万达·湖公馆售楼处绿化

LANDSCAPE DESIGN
售楼处景观

常德古名"武陵",是桃花源出处之地,桃花源所倡导的居住理念,堪称现代慢生活之范本。常德万达售楼处以"慢生活、自由街"为主题,汲取地方文化特色与现代生活方式相结合通过情景雕塑小品及花池坐凳的设计,打造一条有地域文化特色的时尚商业街。使人们在游憩、小酌之间便完成了售楼处与样板间的参观。整体展示区域依据功能总共分为:住宅入口景观区、样板间景观区、售楼处景观区。另加三个特色商业展示区域,为咖啡餐饮展示区、琴行展示区、工艺品商店展示区。住宅入口点缀两棵香樟,与入口门头相得益彰,凸显入口区域的豪华大气;售楼处入口区通过景观灯柱与特色铺装的搭配,保证序列感的同时很好地兼顾了功能性;三大特色商业展示区域通过主题雕塑与小品的点缀,完美诠释了慢生活与自由的真谛。

Changde, called "Wuling" in the ancient time, is the cradle of Taohuayuan (peach blossom garden), the residential concept of which is said to be a model for the modern slow life. Following the theme of "Slow Life, Liberty Street" and combining the local culture characteristics with the contemporary lifestyle, Changde Wanda Sales Office uses situational sculpture accessories, flower bed and bench to build a chic commercial street with regional culture characteristics, where people may visit the sales office and prototype room while wandering around and taking a sip. The whole exhibition area is functionally divided into a residence entrance landscape area, a prototype room landscape area and a sales offices landscape area. Besides, there are three featured commercial exhibition areas, including a café exhibition area, a musical instrument store exhibition area and a craft store exhibition area. The residence entrance furnished with two camphor trees echoes with the entrance gate and delivers a luxurious and grand momentum. The sales office entrance uses landscape lamp post and featured pavement to guarantee both senses of order and function. And the three featured commercial exhibition areas perfectly interpret the truth of slow life and liberty, through theme sculptures and interspersed accessories.

08 常德万达·湖公馆售楼处大厅
09 常德万达·湖公馆售楼处洽谈区
10 常德万达·湖公馆售楼处平面图

INTERIOR DESIGN
售楼处内装

整体风格以"鱼米之乡，富贵花开"为主题。内装设计简约、现代，大厅的主要功能满足展示、接待和营造适合销售的氛围。通过采用中式元素融合现代风格，达到精致不失品位、细节提高质感的效果。中庭上方悬挂水晶吊灯，与LED屏演示画卷、大型电动沙盘形成中心主视觉。顶棚金箔饰面和墙面地面的搭配有效地烘托了主题，体现出了大气的空间氛围。

Adhering to the theme of "A Land of Fish and Rice Fortune of Blooming Flowers", the interior design is simple and stylish. The lobby mainly serves for exhibition and reception, and builds desirable sales atmosphere. After combining the Chinese style elements with the modern style, the exquisite, tasteful and details-oriented quality effect is achieved. The main visual focus is composed of crystal droplight hanging over the atrium, LED screen demonstration picture scroll and large-scale electric building model. Gold leaf facing of ceiling, wall and floor together build a magnificent atmosphere, perfectly foiling the theme.

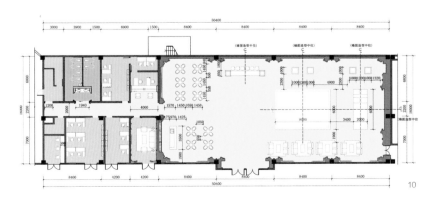

售楼处

06 SHANGHAI QINGPU WANDA MALL EXHIBITION CENTER
上海青浦万达茂展示中心

时间：2015 / 04 /25　　　　**OPENED ON**：25th APRIL, 2015
地点：上海 / 青浦　　　　　**LOCATION**：QINGPU, SHANGHAI
建筑面积：992 平方米　　　**FLOOR AREA**：992M²

01

ARCHITECTURAL DESIGN
售楼处建筑

展示中心作为上海青浦万达茂给予客户的第一印象,要完成引导客户由展示中心去往参观住宅及SOHO实体样板间的任务。展示中心用"桥"为主题,以28种不同色相的彩釉玻璃渐变为主体,结合木纹铝板线条装饰,营造华丽绚烂、庄重典雅的效果,成为沿淀山湖大道的城市标志性建筑物,对上海青浦万达茂的营销工作起到积极的促进作用。

As the first impression of Shanghai Qingpu Wanda Mall felt by clients, the exhibition center should lead clients to the residence and SOHO prototype room. It follows the theme of bridge and largely applies enameled glasses with graduated 28 colors, building gorgeous, solemn and elegant effect, combined with wood grain aluminum plate lines. The exhibition center thus becomes a city landmark building along Dianshanhu Lake Avenue, and positively propels marketing of Shanghai Qingpu Wanda Mall.

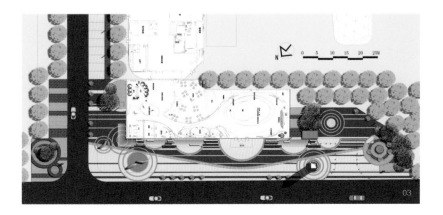

01 上海青浦万达茂展示中心夜景
02 上海青浦万达茂展示中心外立面
03 上海青浦万达茂展示中心总平面图

04 上海青浦万达茂展示中心喷泉
05 上海青浦万达茂展示中心入口
06 上海青浦万达茂展示中心水景

LANDSCAPE DESIGN
售楼处景观

景观提取上海青浦地域特色的"水乡"文化作为项目主题,并结合展示中心建筑设计理念"桥",与景观"水"形成倒影关系,在地表铺装上加以诠释。在水景运用上,采用广场旱喷与建筑边缘跌水的"点"与"线"组合形式。注重互动性设计,通过缓坡式入水及高低错落的石景,丰富了水景的竖向变化与趣味性。提取建筑幕墙的设计语汇及色彩元素,使建筑及室内形成整体统一的设计风格。软景绿化以点状式的小绿化组团为主,选用上海地区本土植物营造多重绿化效果。

The landscape design takes the "water town" culture unique to Shanghai Qingpu as the project theme, and constitutes inverted relation with architectural design idea for the exhibition center "bridge", which is interpreted in floor pavement. The waterscape applies point (dry spray at plaza) "line" (water fall at architectural fringe) combination, and pays keen attention to interactive design such as gentle slope entry and staggered rockscape, enriching vertical change and delight of waterscape. After employing the design language and color element of architectural curtain walls, buildings and interior achieve a generally unified style. The greening of the soft scenery largely applies dotted greening clusters and selects local plants in Shanghai to deliver multiple greening effects.

07

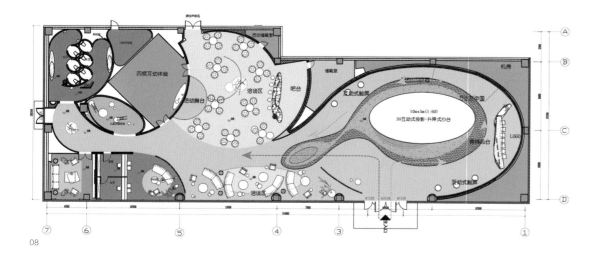

08

07 上海青浦万达茂展示中心大厅
08 上海青浦万达茂展示中心平面图
09 上海青浦万达茂展示中心洽谈区
10 上海青浦万达茂展示中心前厅

INTERIOR DESIGN
售楼处内装

内装设计以上海青浦区具有江南水乡特色"水文化"为依托,以体现"水"给人们带来的灵动、亲切、动静之美;把自然特性融入空间之中,整体空间以"水滴"晶莹剔透的点状形态增强空间动感与生命力,以水波圆圈的重叠关系形成平面构成,丰富界面表现形式;以水纹的弧线提炼动态,以水浪自由的韵律感与邓鹏与地面的各种"水"的姿态相互呼应,让空间整体而顺畅。

The interior design relies on the district's "Water Culture" with Jiangnan Watertown features, to reflect the flexible, intimate and moving beauty of water. To blend the natural characteristics into the space, the whole space adopts water drop-like crystal clear dot shape to attain enhanced dynamism and vitality; it depends on the overlapping ripple circles to achieve plane composition and enrich interface expression; it becomes an integrated whole and smooth, through dynamism made available by water wave arc and harmony between rhythmical water wave and water gesture presented by ceiling and floor.

样板间

01 PROTOTYPE ROOM OF NANPING WANDA CORE MANSION
南平万达中央华城样板间

时间：2015 / 06 / 07　　OPENED ON：7th JUNE, 2015
地点：福建 / 南平　　　　LOCATION：NANPING, FUJIAN PROVINCE
建筑面积：349 平方米　　FLOOR AREA：349M²

01 南平万达中央华城现代欧式风格样板间客厅
02 南平万达中央华城现代欧式风格样板间卧室
03 南平万达中央华城现代欧式风格样板间餐厅

MODERN EUROPEAN STYLE
现代欧式风格样板间

户型为118平方米,用优雅的配饰、丰富的色彩、精美的造型,取得雅致华贵的装饰效果,并表达浓厚的温馨氛围和柔和的古典情调。

With elegant accessories, rich colors and exquisite modeling, the prototype room of 118 square meters is decorated to be elegant and luxuriant, and delivers quite cozy atmosphere and gently classical sentiment.

04

CHINESE STYLE GARDEN HOUSE
中式风格花园洋房样板间

户型为144平方米，设计将现代元素和传统中式元素糅合，配以意境悠远的画作和精致古典的饰品，烘托出中庸平和、高贵典雅的意境。

The prototype room of 144 square meters mixes the modern elements and traditional Chinese elements, and incorporates paintings with profound artistic conception and delicate & classical accessories, presenting peaceful and elegant perception.

05

PART B　COMMERCIAL PROJECTS 商业项目

04 南平万达中央华城现代中式风格花园洋房样板间卧室
05 南平万达中央华城现代中式风格花园洋房样板间餐厅
06 南平万达中央华城现代中式风格花园洋房样板间户型图
07 南平万达中央华城现代中式风格花园洋房样板间客厅

样板间

02 PROTOTYPE ROOM OF NANNING JIANGNAN WANDA PLAZA
南宁江南万达广场样板间

时间：2015 / 09 / 13　　**OPENED ON**：13th SEPTEMBER, 2015
地点：广西 / 南宁　　**LOCATION**：NANNING, GUANGXI ZHUANG AUTONOMOUS REGION
建筑面积：100平方米 –140平方米　　**FLOOR AREA**：100-140M²

01 南宁江南万达广场简欧风格样板间户型图
02 南宁江南万达广场简欧风格样板间餐厅
03 南宁江南万达广场简欧风格样板间客厅
04 南宁江南万达广场简欧风格样板间卧室

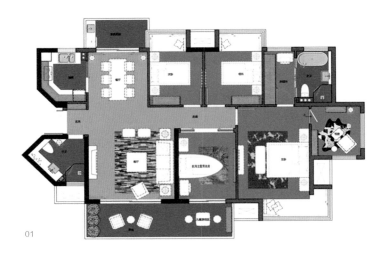

01

02

MODERN EUROPEAN STYLE PROTOTYPE ROOM
简欧风格样板间

本案为140平方米样板间，以融合当地壮锦文化与古典欧式风格而受到普遍的欢迎。这种设计具有优雅、舒适、温馨、华美的特点，将适用于大户型的纯古典欧式风格加以改良，形成适合中等户型并具有强烈地方风格的新简欧风格。

The prototype room of 140 square meters is well received due to its combination between the local Zhuang Brocade culture and classic European style and characterized by being elegant, comfortable, cozy and magnificent. Based on modification of the purely classic European style suitable for large houses, the new Modern European Style full of local characteristics and applicable to medium houses takes shape.

NEW CHINESE STYLE PROTOTYPE ROOM
新中式风格样板间

本样板间为120平方米户型，营造了朴素沉稳、具有浓郁文化底蕴的住宅风格。设计引用古人"朴素而天下莫能与之争美"的精神领悟作为本案的理念。门厅、端景衣柜的山水饰面，客厅简约、朴素的木格栅，简练顶棚以及素雅的软装，给空间带来远离喧闹的感觉。客厅的罗汉床、茶台的紫砂器具、阳台上的围棋、书房里的画卷、卧室里的琵琶——这一切都为居者营造一个沉淀思绪、冲淡浮尘、体会生命本源的"塞外桃源"。

Advocating the idea of Zhuang Zhou that simplicity is the most natural beauty, the prototype room of 120 square meters pursues simple and steady residence style with profound cultural deposits. The landscape facing of entryway and side view wardrobe, simple and plain wooden grille in living room, concise ceiling and elegant soft decoration enable an escape from bustle. And the Arhat bed in living room, purple clay tea set on tea table, the game of go on the balcony, picture scrolls in study and Chinese lute in bedroom all build a heaven of peace, where occupants may meditate, relax and perceive life origin.

样板间

07 PROTOTYPE ROOM OF JIXI WANDA PALACE
鸡西万达华府样板间

时间：2014 / 10 / 31
地点：黑龙江 / 鸡西市
建筑面积：310 平方米

OPENED ON : 31ˢᵗ OCTOBER, 2014
LOCATION : JIXI, HEILONGJIANG PROVINCE
FLOOR AREA : 310M²

NEOCLASSICAL STYLE PROTOTYPE ROOM
新古典式风格样板间

本户型173平方米，温文尔雅的高雅品质、简单流畅的线条、超凡脱俗的吊顶、积淀岁月的咖色主色调、顶级设备的配置等，使得每一处都是温馨低调的奢华，尽情地演绎了高品位的舒适生活格调。

The 173 square meters room pursues gentle and cultivated quality, simple and smooth lines, extraordinary ceiling, age-old coffee color and supreme equipment. All elements fully reveal a low-pitched luxury and high-grade comfortable life style.

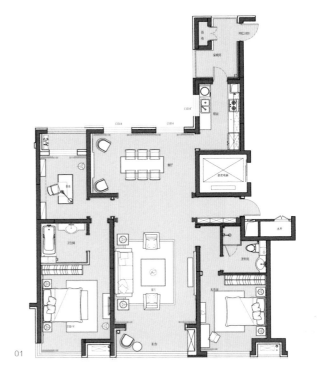

01

02

01 鸡西万达华府新古典风格样板间户型图
02 鸡西万达华府新古典风格样板间餐厅
03 鸡西万达华府新古典风格样板间客厅
04 鸡西万达华府新古典风格样板间卧室

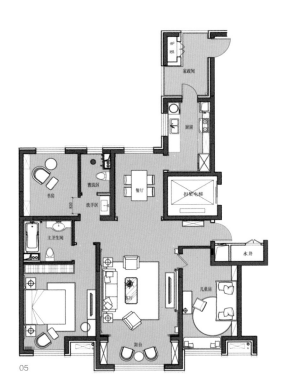
05

06

07

NEW CHINESE STYLE PROTOTYPE ROOM
新中式风格样板间

本户型137平方米，为新中式风格，演绎中式端庄、稳重的精髓，兼具雅致、大气的品位，将现代和传统元素相结合，以现代人的审美打造古典的韵味，运用简约大气的手法为业主营造居所，体现主人品位不凡的生活态度。在空间与造型的处理上，尺度的把握恰到好处。

The prototype room of 137 square meters pursues a new Chinese style. It interprets dignified and steady essence of the Chinese style and elegant and grand quality. It integrates the modern and traditional elements to create classical appeal from the perspective of modern aesthetics. It adopts simple yet magnificent methods to build home for owners, displaying marvelous life attitude bearing by owners. It realizes proper command of scale when dealing with space and modeling.

05 鸡西万达华府新中式风格样板间户型图
06 鸡西万达华府新中式风格样板间书房
07 鸡西万达华府新中式风格样板间客厅
08 鸡西万达华府新中式风格样板间卧室

样板间

08 PROTOTYPE ROOM OF TAIYUAN WANDA PLAZA
太原万达广场样板间

时间：2016 / 05 / 28 **OPENED ON**: 28th MAY, 2016
地点：山西 / 太原 **LOCATION**: TAIYUAN, SHANXI PROVINCE
建筑面积：1326 平方米 **FLOOR AREA**: 1,326M²

EUROPEAN STYLE PROTOTYPE ROOM
欧式风格样板间

本户型230平方米，整体空间比例对称、均衡、大气、空间感强。空间造型优雅精致，很好地融入了欧式风格的精巧与细腻，让人过目不忘。用精致的石膏线条和精美的图案墙纸，以及地面的彩色石材拼花，配以华丽的装饰品，塑造出雍容华贵的效果，体现出太原顶级住宅的气质。

The prototype room of 230 square meters is well-proportioned, balanced, grand and spacious in general. Through incorporation of the exquisite and delicate European style, the elegant and refined space modeling is very impressive. Thanks to fine plaster lines, fancy patterned wallpaper, colorful stone floor pattern and gorgeous ornaments, the room looks both dignified and graceful, and shows the quality of a high-end residence in Taiyuan.

01 太原万达广场欧式风格样板间客厅
02 太原万达广场欧式风格样板间餐厅
03 太原万达广场欧式风格样板间卧室

MODERN STYLE PROTOTYPE ROOM
现代风格样板间

本户型190平方米，用现代的手法配合木色墙面造型，无论是从硬装，还是软装以至细小的饰品来看，都把握住色调的协调，让每处空间看起来和谐雅致，为客户打造出奢华韵味的家。整体空间布局动线流畅、方便，空间线条柔美而富于节奏感。在皮革与金属的"炫酷"装饰品搭配下，混合着时尚的现代气息。

The prototype room of 190 square meters applies the modern methods and wood-colored wall modeling. Coordinated color is found everywhere, from hard decoration, soft decoration to accessories, contributing to harmonious and elegant spaces and rendering clients appealing luxury home. The overall spatial layout enjoys smooth and convenient circulation, and space lines are graceful and rhythmical. With "cool" ornaments made of leather and metal, a fashion modern sense is felt.

04 太原万达广场现代风格样板间餐厅
05 太原万达广场现代风格样板间卧室
06 太原万达广场现代风格样板间客厅

| PART B | COMMERCIAL PROJECTS 商业项目 | 145

01 东莞厚街万达广场实景示范区总平面图
02 东莞厚街万达广场实景示范区景观
03 东莞厚街万达广场实景示范区绿化

PART B | COMMERCIAL PROJECTS 商业项目

04 东莞厚街万达广场实景示范区景观
05 东莞厚街万达广场实景示范区景观设计手绘图
06 东莞厚街万达广场实景示范区景观小品

实景示范区

02 DEMONSTRATION AREA OF FOSHAN SANSHUI WANDA PLAZA
佛山三水万达广场实景示范区

时间：2015 / 10 / 17　　**OPENED ON**：17th OCTOBER, 2015
地点：广东 / 佛山　　　**LOCATION**：FOSHAN, GUANGDONG PROVINCE
占地面积：0.15 公顷　　**LAND AREA**：0.15 HECTARE

01 佛山三水万达广场实景示范区住宅入口
02 佛山三水万达广场实景示范区总平面图

ARCHITECTURAL DESIGN
示范区建筑

住宅立面采用新古典风格，经典的"三段式"，强调对称挺拔向上的气势；干净利落、大块渐成的风格，颇有立体主义的风范。下部采用石材、砖等传统装饰材料体现建筑的端庄、稳重；屋顶采用古典、厚重的装饰风格凸显建筑的典雅、尊贵；建筑形体的凹凸形成挺拔、简洁的竖向线条，显现出建筑高贵而内敛、优雅而沉稳的文化气息。

The residence facade pursues the neoclassical style. Its classical "three-section" highlights symmetrical and towering momentum; it's simple, neat and graduated style quite resembles cubism. Its lower section uses the traditional decoration materials such as stones and bricks to present a dignified and steady building. The roof follows the classical and dignified decoration style to exhibit an elegant and exalted building. Uneven building shape forms straight and simple vertical lines, revealing cultural ambience of building-noble and restrained, elegant and composed.

LANDSCAPE DESIGN
示范区景观

展示区沿主观赏路径设置镜面薄水，配以灵动的涌泉及精致的小品雕塑，既可解决其消防需求，又提供动静相宜、富有格调的景致；沿路布置的林荫休憩平台，提供了休憩及洽谈的舒适空间。样板间出入口植物配置以高大乔木为主，灌木点缀，呈现出富有迎接感且层次丰富的植物造景，营造休闲生态、轻松愉悦的氛围；出入口对景处的景墙、水景在凤凰木的映衬之下，更平添休闲、生态的气息。

Along the main viewing path of the demonstration area, mirror-like water, dynamic fountains and exquisite sculptures are provided to both cater for fire control requirement and offer static, dynamic and stylish scenery. Along the path, shaded rest platforms are also arranged to provide comfortable space for rest and communication. As for the prototype room, its entrance & exit is mainly covered by tall trees and dotted by shrubs, delivering a kind of welcoming and layered plant landscape and building a relaxing and ecological leisure environment. The landscape wall and waterscape of the opposite scenery at the entrance & exit against delonix regia make the ecological leisure atmosphere more prominent.

03 佛山三水万达广场实景示范区绿化
04 佛山三水万达广场实景示范区景观雕塑
05 佛山三水万达广场实景示范区景观

05 DEMONSTRATION AREA OF BOZHOU WANDA PLAZA
亳州万达广场实景示范区

时间：2015 / 05 / 20　　**OPENED ON**：20th MAY, 2015
地点：安徽 / 亳州　　　**LOCATION**：BOZHOU, ANHUI PROVINCE
占地面积：2400 平方米　**LAND AREA**：2,400M²

LANDSCAPE DESIGN
示范区景观

亳州万达广场示范区景观设计倾向于表达优雅的意境，在喧闹的城市空间里呈现宁静感觉和舒缓节奏；通过对空间收放、材质选择、种植色彩搭配，形成丰富而清新的感受。入口林荫道相迎，两侧斑斓的花卉拥簇，营造出温馨的归家之路。曲径通幽的小径与水景相融合，用弧线打破局促空间以及建筑生硬的线条；水景与休憩场地以及景观驻足点贯穿整个中心地带，小桥流水、对景跌水与丰富的种植组团相结合，独景树与花海完美地结合。这一切使居住者沉醉在都市桃花源之中。

The landscape design of the demonstration area strives to deliver elegant artistic conception, and present a tranquil and slow pace amid bustling city space. Its spatial layout, material selection and plants color collocation create colorful yet fresh perception. The welcoming entrance avenue embraced by bright-colored flowers on both sides builds a cozy way home. Witnessing the winding path integrated with waterscape, interruption of cramped space and stiff architectural lines via arc lines, waterscape and place for stop and rest found everywhere in the whole central area, combination of bridges, running water, water fall in the opposite scenery and vast plant clusters, and ideal harmony between flowers and trees, occupants may be intoxicated with the urban Taohuayuan (The Peach Blossom Spring).

01 亳州万达广场实景示范区总平面图
02 亳州万达广场实景示范区水景

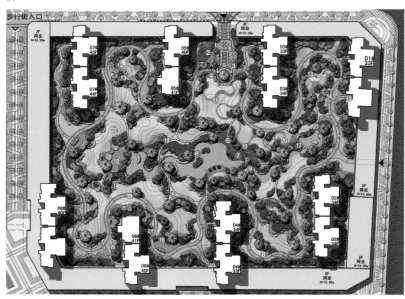
01

02

05

06

03-04 亳州万达广场实景示范区绿化
05-06 亳州万达广场实景示范区花园
07 亳州万达广场实景示范区木桥

实景示范区

06 DEMONSTRATION AREA OF SHANGRAO WANDA PLAZA
上饶万达广场实景示范区

时间：2015 / 05 / 31　　OPENED ON：31st MAY, 2015
地点：江西 / 上饶　　　LOCATION：SHANGRAO, JIANGXI PROVINCE
占地面积：0.62 公顷　　LAND AREA：0.62 HECTARE

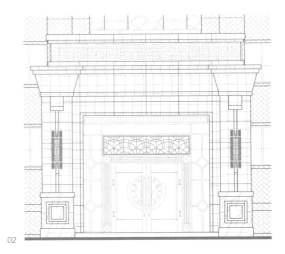

ARCHITECTURAL DESIGN
示范区建筑

上饶市婺源县以其独特的徽州古韵，以及春天油菜花绽放的美景被誉为"中国最美丽的乡村"。上饶万达广场位于上饶市中心，信江江畔，在建筑设计上以徽派风格为"底色"，形成总体上临江屹立、飞檐重叠的天际形象。

Wuyuan County of Shangrao City is honored as the "Most Beautiful Countryside in China" for its unique Huizhou style and blooming rape flowers in spring. Shangrao Wanda Plaza is located in downtown of Shangrao and on the Xin River bank. Its architectural design is based on the Hui style, presenting a skyline image erecting by the river with overlapped cornice.

01 上饶万达广场实景示范区总平面图
02 上饶万达广场实景示范区住宅入口立面图
03 上饶万达广场实景示范区住宅立面图
04 上饶万达广场实景示范区住宅入口

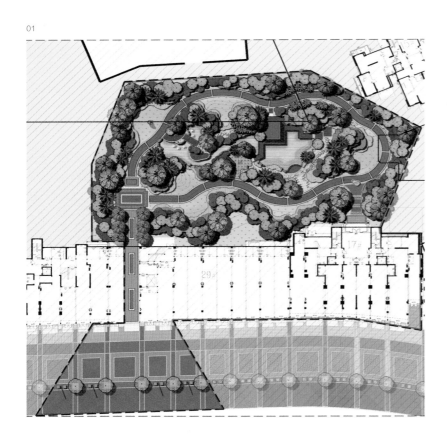

LANDSCAPE DESIGN
示范区景观

景观设计沿用"婺源风韵"的文化主题，通过新中式的表现手法，突出上饶万达广场的地域文化性和独特性。通过对空间、场所的收放开合，铺装拼花的精心选择以及种植的色彩搭配，体现示范区内部空间大气和精致的变化。曲径通幽的小径与水景相融合，用柔美的曲线打破生硬的线条，水景与休憩场地以及各个景观驻足点贯穿整个中心地带，吸引居住者进入室外享用园林。

The landscape design sticks to the cultural theme of "Wuyuan Charm", and applies the new Chinese style expression technique to highlight the regional culture and peculiarity of Shangrao Wanda Plaza. Through the flexible layout of spaces and places, careful selection of pavement pattern and color collocation of plants, magnificent and delicate interior change of demonstration area is touched. Through winding path integrated with waterscape, interruption of stiff lines via gentle curves, waterscape and place for stop and rest found everywhere in the whole central area, occupants are drawn outdoors to appreciate the garden scenery.

05 上饶万达广场实景示范区景观亭
06 上饶万达广场实景示范区景观设计手绘图
07 上饶万达广场实景示范区水景
08 上饶万达广场实景示范区绿化

PART **B** | COMMERCIAL PROJECTS
商业项目

06

03 HEFEI WANDA CULTURAL TOURISM CITY
合肥万达城

时间：2015 / 06 / 30
地点：安徽 / 合肥
占地面积：18.9公顷
建筑面积：76.0万平方米

OPENED ON: 30th JUNE, 2015
LOCATION: HEFEI, ANHUI PROVINCE
LAND AREA: 18.9 HECTARES
FLOOR AREA: 760,000M²

PROJECT OVERVIEW
项目概况

合肥万达城位于安徽省合肥市滨湖新区巢湖岸边。其中一期规划用地面积为18.9公顷，南临成都路，北至云谷路，东起澳门路，西至华山路。一期住宅设计传承古典与现代结合的气质，营建皇家级别的园林，打造合肥顶级高贵住区。

Hefei Wanda Cultural Tourism City is located in Chaohu Lake bank of Binhu New District, Hefei, and faces Chengdu Road to the south, Yungu Road to the north, Macao Road to the east and Huashan Road to the west. Covering a planned site area of 18.9 hectares, the Phase I residence pursues both the classical and modern styles to construct a royal garden and high-end residential area.

01 合肥万达城总平面图
02 合肥万达城夜景
03 合肥万达城住宅入口

ARCHITECTURAL PLANNING
建筑规划

合肥万达城用开放社区的理念，用商业街将小区分成六个合理的组团；采用大栋距、低密度的设计手法，南北最大栋距达92.7米。在满足日照和间距的情况下，房型均为大面宽设计，充分利用面宽资源，增加南面的房间数量。

规划方面，采用较为方正的规划格局，有利于地下车库的经济排量；总图设计中减少大面积水景，采用点状浅水池，减少造价。

建筑方面，平面取消曲线，减少体形系数，降低能耗；户型结构合理、方正，适当减少开窗面积及玻璃栏板的面积，适当简化建筑细部，以降低施工难度。

Complying with the open community idea, Hefei Wanda Cultural Tourism City rationally divides the community into six clusters. Guided by the design method of high building distance and low density, the north-south distance reaches 92.7m to the maximum. Houses all have large face width with qualified daylighting and distance, so rooms in the south are provided more to make full use of the face width resource.

Relatively square and upright layout is adopted as planned to benefit economical displacement of the underground garage. In the master plan design, the large-scale waterscape is replaced by dotted shallow pool to reduce cost.

In terms of architecture, plane curves are reduced to lower shape coefficient and save energy. While ensuring rational and orderly house layout, area of window and glass baluster is properly decreased, and the architectural details are appropriately simplified to reduce the construction difficulty.

LANDSCAPE DESIGN
景观

设计采用中轴对称式空间，体现尊贵之感，并采用中西合璧的造园手法，塑造出工整典雅的形象。设计注重如下四个方面：第一，"以人为本"——创造舒适宜人的可人环境；第二，"以绿为主"——最大限度提高绿视率，体现自然生态；第三，"因地制宜"——体现"适地适树"、"适景适树"的原则；第四，"崇尚自然"——寻求人与自然的和谐。

The landscape design makes the space in centraxonial symmetry to show dignified quality, and follows the garden order that integrates the Chinese and Western elements to create a neat and elegant image. The design pays attention to the following four aspects: first, "human oriented", i.e. to build comfortable environment; second, "green oriented", i.e. to show natural ecology by maximizing green looking ratio; third, "location oriented", i.e. to follow the principle of matching the site with trees and matching the scenery with trees; fourth, "nature-admiring", i.e. to seek harmony between man and nature.

04 合肥万达城绿化
05 合肥万达城景观
06 合肥万达城水景

MODERN STYLE PROTOTYPE ROOM
现代风格样板间

内装非常注重居室空间的布局与使用功能的完美结合；软装色彩跳跃及高纯色彩的运用，以黄色、红色、蓝色的色彩搭配，大胆而灵活，不单是对现代风格家居的遵循，也是个性的展示，突出时尚与潮流，营造出浓郁现代气息、舒适、优雅的生活。

The interior design pays keen attention to an ideal combination of the spatial layout and use function. The soft decoration boldly and flexibly uses alternated colors and high pure colors, such as yellow, red and blue, which not only complies with the designed modern style but displays individuality. In this way, fashion and trend are highlighted, and rich modern sense and comfortable & elegant life are felt.

07 合肥万达城现代风格样板间户型图
08 合肥万达城现代风格样板间客厅
09-10 合肥万达城现代风格样板间卧室

04 WUHAN CENTRAL CULTURE DISTRICT
武汉中央文化区

时间：2015 / 10 / 26
地点：湖北 / 武汉
占地面积：1.45 公顷
建筑面积：39.0 万平方米

OPENED ON : 26th OCTOBER, 2015
LOCATION : WUHAN, HUBEI PROVINCE
LAND AREA : 1.45 HECTARES
FLOOR AREA : 390,000M²

PROJECT OVERVIEW
项目概况

武汉中央文化区御湖世家（K9地块）位于武汉市武昌区东沙大道与沙湖大道交叉口，总建筑面积39万平方米，其中地上建筑面积31万平方米，地下建筑面积8万平方米。

Wuhan Central Culture District Palais Du Lac Royal (Parcel K9) is located in Wuhan's Wuchang District, the intersection between Dongsha Avenue and Shahu Avenue, with gross floor area of 390,000 square meters, including 310,000 square meters aboveground and 80,000 square meters underground.

01 武汉中央文化区总平面图
02 武汉中央文化区住宅入口

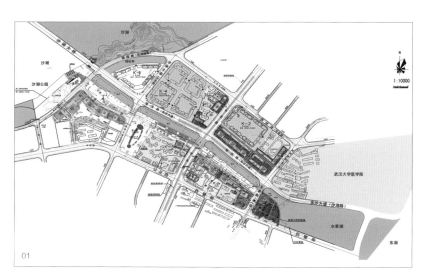

ARCHITECTURAL PLANNING
建筑规划

御湖世家是由四栋超高层住宅及两层商业裙房围合而成一个三角形地块。地块北面临沙湖，景观视野开阔，为一线观湖豪宅。K9地块的超高层住宅与楚河对岸的K6、SK7地块上高耸挺拔的建筑群，一同构筑了大桥两侧重要的地标性建筑。

建筑外立面采用ArtDeco风格，既注重功能性，又不失装饰性。线条在竖向原则上灵活运用，通过重复、对称、渐变，给人以华贵高尚的尊贵感。立面采用对称性和"三段式"的构图原则。顶部层层收缩呈阶梯状，强调垂直线条的高耸感。立面材质以浅色为主，多采用米黄、咖啡色的天然石材，与玻璃幕墙相结合。

Palais Du Lac Royal is a triangle site enclosed by four super high-rise residences and a two-floor commercial podium. Facing Shahu Lake and enjoying open view; it is a luxury residential project with a full-view of the river. Super high-rise residence on Parcel K9 and towering building group erecting on the other side of the Chu River together constitute an important landmark on both sides of the bridge.

The facade in the ArtDeco style accents both function and decorative effect. The flexible vertical lines deliver elegant dignity through repetition, symmetry and gradation treatments. Following the "three-section" and symmetrical facade composition principle, the ladder like layered contraction at the top highlights the towering vertical lines. The facade largely applies light-colored materials, such as beige and brown natural stones, to combine with the glass curtain wall.

03 武汉中央文化区示范区入口
04 武汉中央文化区示范区建筑外立面

LANDSCAPE DESIGN
景观

御湖世家景观的理念为"三湖（东湖、沙湖、楚河）围绕"，形成一个半岛的地理形态，取得独到的自然文化优势，因此以"汉水之滴"作为创作主题。项目以简欧的园林风格结合自然元素，力图打造一个既具有独特人文景观又实现精致生活的高端住宅环境。将建筑风格加以提炼，融合到景观中，强调入口轴线景观，将不同功能、形态的区域联系成紧密的整体，形成紧凑而丰富的序列空间。运用中国传统园林中对景、夹景、借景等手法，使景观小品、植物、地形、水景和谐共处，营造出高贵典雅的园林氛围。

With view to the idea of "Three Lakes Around" (i.e. East Lake, Shahu Lake and Chu River) to form a peninsula geographical appearance and gain unique natural and cultural superiority, the landscape design applies the theme of "Drop of the Han River". Through a combination between the Modern European Style garden and natural elements, the project strives to build a high-end residential environment housing distinctive cultural landscape and pursuing delicate life. Through integrating the architectural style into landscape, the project accents entrance axis landscape. Through linking areas of different functions and forming into a whole, a compact, colorful and sequential space takes shape. Through applying methods of the traditional Chinese garden (opposite scenery, vista line and borrowed scenery) to achieve harmony among landscape pieces, plants, terrain and waterscape, noble and elegant garden atmosphere is felt.

05 武汉中央文化区示范区中央水景
06 武汉中央文化区示范区景观花园
07 武汉中央文化区示范区景观连廊

08 武汉中央文化区法式风格样板间门厅
09 武汉中央文化区法式风格样板间客厅
10 武汉中央文化区法式风格样板间户型图
11 武汉中央文化区法式风格样板间卫生间

LANDSCAPE DESIGN
景观

景观设计注重处理好人、建筑与环境"三者"的协调关系。居住环境采用"园林化"的设计理念,为居民放松、休憩、活动和交往提供了有益的空间。借鉴中国传统园林中对景、夹景、借景等手法,使园林小品、植物、建筑交相呼应;利用地形高差,建造错落有致的精致水景,体现高贵典雅的园林氛围;通过花瓣造型主题、设计环形广场和台地花境,使得背景的"溪流林地"成为中轴视线焦点,强调入口轴线景观,将不同功能、形态的区域连接成紧凑而丰富的序列空间。

The landscape design is concerned about coordinated relation among people, buildings and environment. The living environment design pursues a "garden" idea to provide occupants space for relaxation, rest, activity and interaction. Through applying methods of the traditional Chinese garden (the opposite scenery, vista line and borrowed scenery), harmony among garden accessories, plants, terrain and buildings is achieved. Through the well-arranged exquisite waterscape built by taking advantage of the terrain height difference, noble and elegant garden atmosphere is felt. Through a petal theme, the circular plaza and terrace flower border, stream & woodland originally served as background becomes visual focus on the center axis, with an emphasis on the entrance axis landscape. Through linking areas of different functions and forms into a whole, a compact, colorful and sequential space takes shape.

04 南昌万达城示范区水景
05-06 南昌万达城示范区绿化
07 南昌万达城示范区景观

08-09 南昌万达城地中海风格样板间卧室
10 南昌万达城地中海风格样板间客厅
11 南昌万达城地中海风格样板间户型图

MEDITERRANEAN STYLE PROTOTYPE ROOM
地中海风格样板间

本样板间建筑面积95平方米，整体空间在布局上秉承惯有的简洁、大气、现代之风格，并且通过材料与软装的结合，使空间简约而不简单，适合现代人的生活需求与意识审美。空间以清新简约、自由奔放的湖蓝色和淡青绿色为主，带有现代欧式韵味的家具，搭配舒适柔软的布艺，使人仿佛置身戛纳海边度假的悠闲。

The 95-square-meter prototype room still pursues the invariable concise, grand and modern style in its overall space layout. Besides, by combining materials with soft decoration, the space is concise yet not simple, which lives up to life demands and aesthetic consciousness in modern times. The space design gives priority to lake blue and light blue-green colors and arranges the modern European style furniture, comfortable and soft fabrics, offering the experience of enjoying holiday by Cannes seaside.

12 南昌万达城禅意风格样板间客厅
13 南昌万达城禅意风格样板间户型图
14 南昌万达城禅意风格样板间书房

ZEN STYLE PROTOTYPE ROOM
禅意风格样板间

本样板间建筑面积115平方米，大多以纯天然的柚木为材质，纯手工制作而成，取材天然，带着几分拙朴，原汁原味；参差不齐的柚木没有任何修饰，却仿佛藏着无数的禅机。实木家具、棉麻材质的布艺以及藤条编制的装饰品，将各种家具包括饰品的颜色控制在棕色或咖啡色系范围内，再用白色墙面调和，营造出东方人特有的禅雅意境。

The 115 square meter prototype room is largely furnished with handcrafted natural teak, simple, unadorned yet full of Buddhist Allegory. Solid wood furniture, cotton & linen fabrics and cane braided ornaments in brown or coffee color series against overwhelmingly white color together build a Zen style unique to the Easterner.

07 HARBIN WANDA CULTURAL TOURISM CITY
哈尔滨万达城

时间：2015 / 10 / 30
地点：黑龙江 / 哈尔滨
占地面积：30.0公顷
建筑面积：90.0万平方米

OPENED ON : 30th OCTOBER, 2015
LOCATION : HARBIN, HEILONGJIANG PROVINCE
LAND AREA : 30.0 HECTARES
FLOOR AREA : 900,000M²

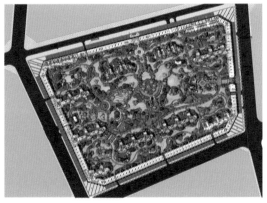

PROJECT OVERVIEW
项目概况

哈尔滨万达城3号地块位于万达城的西侧，为高层区域。其地上建筑面积57.3万平方米，包括1栋100米的高层公寓、4栋150米的超高层住宅、6栋100米的高层住宅和沿街2层的商业裙房。绿地率36.96%，总绿化面积6.5万平方米。

The Parcel 3# is located in the west of Harbin Wanda Cultural Tourism City and it is a high-rise area with aboveground gross floor area of 573,000 square meters. The area houses one 100m high-rise apartment, four 150m super high-rise residences and six 100m high-rise residences and a two-floor commercial podium along the street, with greening rate being 36.96% and total green area being 65,000 square meters.

01 哈尔滨万达城总平面图
02 哈尔滨万达城住宅外立面

ARCHITECTURAL PLANNING
建筑规划

哈尔滨万达城设计风格为简约欧式，营造典雅、自然、高贵的气质。浪漫的情调是本案的主题，简欧风格继承了传统欧式风格的特点，在设计上追求空间变化的对称性和形体变化的层次感，给人营造一种整列、大气的仪式感和尊贵感。

Harbin Wanda Cultural Tourism City pursues the Modern European Style to render elegant, natural and exalted quality. Following the theme of romanticism, the style inherits features of the traditional European style that is to seek symmetrical spatial changes and layered shape variation, presenting an orderly and magnificent ritual sense and an exalted sense.

03

LANDSCAPE DESIGN
景观

哈尔滨万达城3号地块景观，定位为自然简欧风格。走进园区，仿佛身处欧式的花园中——开阔的空间场地、蜿蜒开合的园区道路、大气恢宏的中心台地广场、别具特色的儿童乐园，以及丰富的植被绿化——这一切都与这里的建筑相映生辉。

The Parcel 3's landscape pursues the natural Modern European Style. Walking into the site, one may appreciate the European style garden landscape: a spacious venue, a winding site road, a splendid central terrace plaza, a featured kids place and vast plants greening, all echoing with building thereof.

03 哈尔滨万达城示范区景观
04 哈尔滨万达城示范区花坛

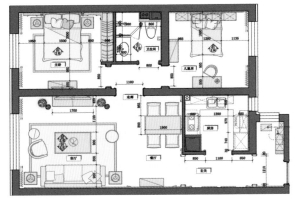

07

05 哈尔滨万达城田园风格样板间户型图
06 哈尔滨万达城田园风格样板间客厅
07 哈尔滨万达城田园风格样板间卧室

COUNTRY STYLE PROTOTYPE ROOM
田园风格样板间

本户型建筑面积107平方米，装修以纯天然的柚木为主材，纯手工制作而成，表达贴近自然、向往自然的倾向。采用田园风格的家具，色彩丰富，色调饱和，展现自然气息。设计以具有亲和力材质、体现田园风情家具及自然风格的色调组合，营造受到都市人喜爱的生活空间。

The 107-square-meter prototype room is largely furnished with handcrafted natural teak to show its preference of nature. The country style furniture is colorful and bright, well displaying the natural atmosphere. Thanks to these amiable materials, country style furniture and natural colors combination, living space favored by city dwellers is constructed.

MODERN STYLE PROTOTYPE ROOM
现代风格样板间

本户型建筑面积127平方米，以平面和直线造型造成视觉的通透感，营造简洁的风格。顶面、墙面共同组成了房间的主色调，灰白相间的电视背景墙，与米黄色的沙发相互映衬、自然过渡。黑色的电视机与地毯相互呼应，和米黄色的地面一起与主色调形成强烈的对比。整个房间时尚而具有品位，与主人的身份非常契合。

The 127-square-meter room is visually spacious and simple with plane and straight line modeling. The ceiling and wall color dominates the room; the grey-white TV background wall and beige sofa set off each other to form a natural transition; the black TV echoing with the carpet and beige floor together constitute a dramatic contrast with primary color. The whole room is stylish and tasteful, and a perfect match with status of the owner.

08 哈尔滨万达城现代风格样板间户型图
09 哈尔滨万达城现代风格样板间餐厅
10 哈尔滨万达城现代风格样板间客厅

08

06 西双版纳万达城示范区喷泉
07 西双版纳万达城示范区景观大道
08 西双版纳万达城示范区木桥
09 西双版纳万达城示范区绿化
10 西双版纳万达城示范区水景

09

10

11

INTERIOR DESIGN
内装

充分利用版纳当地的自然资源，结合内部装修进行精心而富有创意的设计，体现"生活就是度假"的新居住方式。萃取了禅意、东南亚、地中海三种装修风格对室内空间进行精心打造，通过平面优化，使其功能强大尺度舒适；通过院落组合，使前中后院尽享阳光；通过空间优化，使空间细节精雕细刻；通过空间优化，使空间百变附加值高；通过精挑细选，使内装品质更高。

Through thorough utilization of the local natural resources, the interior decoration is subject to elaborate and innovative design, demonstrating a new way of living: life is a vacation. Through application of three decoration styles (the Zen style, Southeast Asian style and Mediterranean style) and plane optimization, the interior space enjoys powerful functions and comfortable scale. The courtyard combination makes front, middle and back courtyards well lighted. Space optimization impels painstaking details treatment and enhances space flexibility. And meticulous selection contributes to an improved interior decoration quality.

12

11-12 西双版纳万达城样板间客厅
13 西双版纳万达城样板间花园
14 西双版纳万达城样板间露台

10 CHENGDU WANDA CULTURAL TOURISM CITY EXHIBITION CENTER
成都万达城展示中心

时间：2015 / 09 / 23
地点：四川 / 成都
占地面积：1.5 公顷
建筑面积：5500 平方米

OPENED ON: 23rd SEPTEMBER, 2015
LOCATION: CHENGDU, SICHUAN PROVINCE
LAND AREA: 1.5 HECTARES
FLOOR AREA: 5,500M²

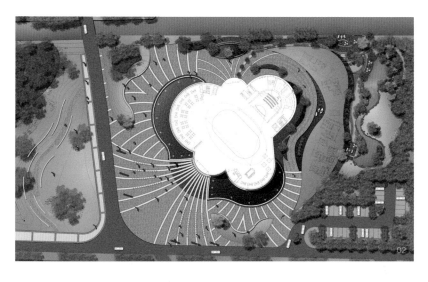

PROJECT OVERVIEW
项目概况

成都万达城是万达集团的第十个"文化旅游城",位于成都下属的都江堰市,总用地320公顷,地上总建筑面积约424万平方米;其中展示中心占地1.5公顷,地上总建筑面积5000平方米。展示中心位于整个项目用地的中部,其通达性、展示性均较好。展示中心建筑以成都市市花"芙蓉花"为主题,其造型大气新颖,优雅美观,成为成都万达城的标志性建筑。

As the tenth Wanda Group's cultural tourism city, Chengdu Wanda Cultural Tourism City is located in Dujiangyan City, Chengdu, and occupies 320 hectares with gross floor area of 4.24 million square meters, among which the exhibition center occupies 1.5 hectares with gross aboveground floor area of 5,000 square meters. Located in the middle of the site, the exhibition center enjoys favorable accessibility and exhibition function. Taking "Lotus", the city flower of Chengdu, as the theme, the magnificent, novel, elegant and aesthetic exhibition center is set to be a landmark building of Chengdu Wanda Cultural Tourism City.

01 成都万达城展示中心鸟瞰图
02 成都万达城展示中心总平面图
03 成都万达城展示中心夜景

04

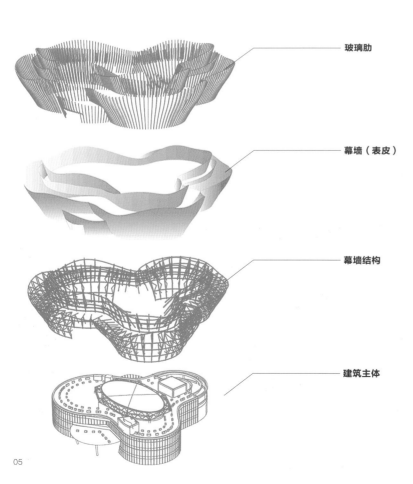

玻璃肋

幕墙（表皮）

幕墙结构

建筑主体

04 成都万达城展示中心外立面
05 成都万达城展示中心设计意向图
06 成都万达城展示中心夜景
07 成都万达城展示中心幕墙特写

05

ARCHITECTURAL DESIGN
展示中心建筑

规划层面，展示中心采用整体设计的理念，将展示中心和其南侧的样板示范区统一规划，采用"红花绿叶，交相辉映"的设计概念，将建筑与空间有机地组织起来，形成了丰富动人的建筑群体效果。在建筑层面，主要在三个方面进行了重点处理和大胆创新。第一，在建筑形态方面，从抽象的、具象的，规则的、不规则的，集中的、展开的等多种方案中，选定形态最接近于"芙蓉花"本体姿态的方案。第二，在建筑幕墙的色彩和质感方面，为了尽可能完美再现"芙蓉花"花瓣的特点，最终选定了"数码打印玻璃"这种新工艺玻璃。第三，在夜景照明方面进行了大胆的尝试，将夜景的主题概念用动画的形式表现出来，强化建筑的形态、突出"芙蓉花"色彩变化的美感及川蜀文化的底蕴。

In terms of planning, the overall design is adopted to apply integrated planning of the exhibition center and its southern demonstration area. The design concept of the "green leaves setting off the red flower" is adopted to organically arrange buildings and space, presenting rich and appealing building group. In terms of architecture, the design emphasis and bold innovation are concentrated on three aspects. First, when dealing with the building shape, the design scheme with shape mostly resembling lotus flower is selected from abstract, concrete, regular, irregular, folded and unfolded shape design schemes. Second, when dealing with color and texture of the curtain wall, new art glass-digital printed glass is a final choice, to perfectly reproduce lotus petal features. Third, when dealing with the nightscape, audacious attempt is made on nightscape lighting-delivering nightscape theme in the form of animation, which consolidates building shape, and highlights color change beauty of lotus and details of the Sichuan culture.

LANDSCAPE DESIGN
展示中心景观

景观设计精心挑选材质，充分利用倒影或人工刻画"芙蓉花"的花韵，使景观映衬出建筑"花开怒放"的姿态。这种处理将建筑的花瓣肌理延展到广场铺装，从而表现出建筑从空间到平面犹如"芙蓉花朵"般层层绽放、交相呼应的效果。根据营销路线安排不同的景观主题，提升场地气质，使客户感受到"百印繁花"的景观意境。

The landscape design is particular about materials and fully utilizes reflection or manual work to depict lotus charm, so that landscape may set off "blooming flowers" gesture of buildings. With the same treatment, petal texture of buildings is extended to the plaza pavement, so that the effect of lotus flower-like blossom in layers is found from space to plane in buildings. Besides, the landscape themes vary with marketing route to enhance site quality and deliver artistic conception of "a mass of flowers" to clients.

PART C | TOURISM PROJECTS
文旅项目

08 成都万达城展示中心鸟瞰图
09-10 成都万达城展示中心景观
11 成都万达城展示中心景观设计手绘图

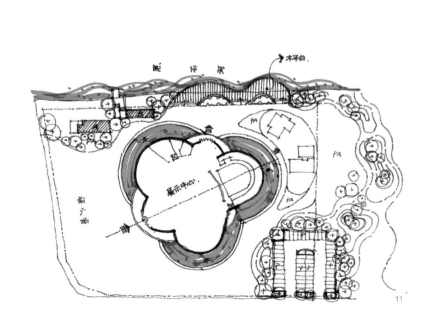

工作内容职责，所有重要内容都反映在合同的"18条核心条款"中。万达"设计总包"包含的设计工作内容包括（但不限于）土建施工图、外立面、幕墙、内装、景观、夜景照明、弱电智能化、导向标识、车库动线等分项设计。为了保证项目的顺利进行，要求"设计总包"单位必须是"万达设计供方库"内的土建施工图设计单位，各分项设计单位必须在"分项万达供方库"内选择。这样的规定要求所有参与设计单位都是万达的合格供方，对万达项目有较深的了解和经验，从而保证了项目的设计成果质量。

《设计总包操作手册》是辅助"设计总包"进行项目实操的指引文件，主要内容分为五章：管理体系、权责界面、管理流程、考核评估、协调仲裁。该"手册"更加系统地明确了各方的"权责"划分及工作界面划分，规范了设计管理的程序及流程，严肃了供方考核制度，同时建立健全了协调与仲裁机制。

二、交流培训

随着集团开发模式、设计总包管理制度的不断创新，制度标准随之更新调整。如何及时有效地把这些变化及时传达给"供方"，实现无缝对接，成为设计管控面临的新问题。

从2015年下半年开始，在"设计总包"管理文件编制过程中，设计中心就邀请"设计总包"单位进行研讨，对管理模式、管理文件等进行了多次论证和修订。在"总包"管理文件"初稿"编制完成之后，设计中心组织开展了多轮的"总包"管理制度培训工作，更加系统地介绍了集团总部、项目公司、总包单位、分包单位的相互关系和工作流程，为"设计总包"模式的实施奠定了基础。

2015年底"设计总包"借助微信平台组建"供方联盟"，大幅提高了交流沟通效率，收效较大，后期逐步成立各专业技术联盟。通过组织一系列的技术研讨活动，借力"设计总包"专业技术领头人力量为集团"技术要点"进步完善提供宝贵意见。设计中心按月编制《设计中心技术管理通报》，及时分享集团新标准、新管控思路，同时为"设计总包"之间搭建设计及管控经验分享平台。

有组织的管理培训和技术管理通报及专项技术联盟管理模式的综合运用，加强了万达与供方在管控思路、方法及信息的有效匹配，提高了效率，并且加强了各供方间的交流，更加有效地整合了各单位的技术资源，实现了提高"设计总包"单位管理能力，与万达共担共赢的良性循环。

engineering construction drawings, facade, curtain wall, interior, landscape, nightscape lighting, extra low voltage intelligence, guiding sign and garage circulation. To ensure the smooth progress of the projects, the main design contractor shall be a civil engineering construction drawings design companies included in Wanda's Design Supplier List and the Sub-contractors shall be selected from Wanda's Sub-items Design Suppliers List. Under such requirement, design companies involved in the projects are all qualified design suppliers acknowledged by Wanda Group and have a better understanding and rich experience for Wanda's projects, thereby guaranteeing the quality of the design results of the projects.

Design Turnkey Operation Manual is a guidance document that assists the Main Design Contractor during the project implementation phase. It has five chapters, which are management system, rights and liabilities, management process, assessment and evaluation and coordination and arbitration. This Manual systematically defines the division of rights and liabilities as well as work duties of each party, standardizes the procedures of design management, settles the strict assessment system for the supplier and establishes and steadily improves the coordination and arbitration mechanism at the same time.

II. COMMUNICATION AND TRAINING

With the continuous innovation of the Group's development model and Design Turnkey Management Model, the standards for the models shall be updated and regulated accordingly. How to deliver these changes to the suppliers in timely and effective manner for seamless connection has become a new problem facing design control work.

From the second half of 2015, in the process of compiling management documents for Design Turnkey Model, the Design Center has invited Main Design Contractors to attend seminars and has argued on and revised the management model and management documents time and again. After the first draft of the Design Turnkey management documents is compiled, the Design Center has organized several rounds of management system training to systematically introduce the interrelations among the Wanda Group headquarters, Project Company, Main Contractor and Sub-contractors and work flow, laying a foundation for the implementation of Design Turnkey Contract Model.

At the end of 2015, the Main Design Contractors set up a "Wechat" Group named Suppliers Alliance. With the help of this platform, the communication efficiency is greatly improved, yielding a satisfying result. Later, the Technical Alliances of each discipline are gradually established. With the organization of a series of technology seminars and with the strength of specialized technical leaders of Design Turnkey Contract Model, valuable suggestions have been proposed for the progress and perfection of the Group's "Technical Essentials". In addition, the Design Center has issued Technical Management Bulletin of Design Center monthly for timely circulation of the Group's new standards and new control ideas. This bulletin is also served as a platform for Main Design Contractors to share their experience in design and control.

With the help of an integrated application of organized management training, Technical Management Bulletin and the management model of special Technical Alliances, the ideas, methods and information concerning control are effectively communicated between Wanda Group and the suppliers, thereby improving work efficiency; the communication

三、实践考核

万达集团于2015年下半年开始在销售物业项目中试行"设计总包"制度。柳南项目售楼处作为第一个"试点总包"项目在2015年10月精彩开放，获得了巨大成功。持有物业于2016年1月正式实施了"设计总包"的招标试点工作。最初先有新乡项目、衡阳项目、余杭项目、溧阳项目等四个试点项目，然后在此基础上全面展开，现在已有二十多个项目实施了"设计总包"模式。针对在试点项目中出现的各种问题，设计中心多次组织"总包"单位召开"总包"交流会，对"设计总包"制度管理文件进行了多次修订，为以后的"设计总包"项目的顺利实施扫清了障碍。

在"设计总包"项目全面开展之后，设计中心将在试点项目的阶段成果完成后进行复盘总结工作，将项目中实际遇到的问题和经验教训加以总结，并对项目公司和"设计总包"单位进行培训。

设计中心负责对"设计总包"单位进行考核评估。"总包"制度要求对"总包"单位按月评分，并进行年度累计评分。"设计总包"的考核单位为设计中心、项目公司以及设计分包，评分分为"技术评分"和"服务评分"。对"总包"单位的每项罚分都必须有明确的事项说明，以及"总包"单位的签字确认，确保评分的"公正、公开和公平"。年度累积之后，评出优秀供方，并给予相应的奖励。

四、未来展望

2017年万达将全面实施"BIM总发包模式"，"业主"、"设计总包"、"工程总包"和"监理单位"通过"四方平台"进行工作对接。"设计总包模式"是"BIM总发包模式"实施的前提条件，为万达向"BIM总发包管理模式"转变奠定了坚实的基础。设计中心将根据新的项目管控要求，进一步完善"设计总包"管理制度和管理文件，并在持有物业"设计总包"成功经验的基础上，在销售物业全面推行"设计总包模式"。希望通过万达实践，为中国建设行业的"设计总包"起到推动作用。

between suppliers are strengthened as well, thereby effectively integrating the technical resources of each company; the management ability of the main design contractors is improved, thereby forming a virtuous circle where the contractor share responsibilities and success with Wanda.

III. PRACTICE ASSESSMENT

Wanda Group started a trial implementation of the Design Turnkey System on projects of properties for sale in the second half of 2015. As the first pilot project applying the Design Turnkey Model, the sales office of Liunan Project opened in October 2015 and achieved a huge success. The pilot bidding for properties for holding projects adopting the Design Turnkey Model was officially started in January 2016. It started with four pilot projects, which are the Xinxiang Project, the Hengyang Project, the Yuhang Project and the Liyang Project, based on which the Design Turnkey Model was implemented in full swing. Now, the Design Turnkey Model has been applied to more than twenty projects. Against those problems found in the pilot projects, the Design Center has organized the main contractors to hold several seminars concerning this model, and has repeatedly revised management documents of the Design Turnkey System, sweeping off obstacles for the future smooth implementation of the Design Turnkey Projects.

Following the full implementation of the Design Turnkey Projects, the Design Center will check and summarize the phased achievements of these pilot projects, summarize the problems encountered and lessons drawn during the project implementation, and conduct training for project company and Main Design Contractors.

The Design Center takes charge of the assessment and evaluation on Main Design Contractors. It is stipulated by the Turnkey System that Main Contractors shall be graded monthly and grades shall be aggregated annually. The companies responsible for assessing Main Design Contractors include the Design Center, the Project Company and Design Subcontractors. Grading shall be carried out for both techniques and services. Specific explanation shall be given for each penalty point deducted for the Main Contractors with the signature confirmation of the Main Contractors to ensure the impartiality, publicity and fairness of the grading. Based on the grades aggregated annually, the excellent supplies are awarded and rewarded.

IV. PROSPECTS

In 2017, Wanda Group will put the "BIM Turnkey Contract Model" in full operation, under which four-party (namely the client, the main design contractor, the main project contractor and the supervisors) platform will be used for work coordination. Being a precondition for the implementation of the BIM Turnkey Contract Model, the Design Turnkey Model has laid a solid foundation for Wanda's transformation towards the BIM Turnkey Contract Model. The Design Center will further improve the management system and management documents of the Design Turnkey Model as per the new project management and control requirements. In addition, on the basis of the successful experience of implementing the Design Turnkey Model for the properties for holding, the Design Turnkey Model shall be put into full implementation for properties for sale. It is hoped that the practices made by Wanda can boost the development of the Design Turnkey Model in Chinese construction industry.

PROMOTING "BIM TURNKEY CONTRACT MANAGEMENT MODEL", PIONEERING INNOVATION ON DESIGN MANAGEMENT
推行"BIM 总发包管理模式",革命性创新设计管理

万达商业地产设计中心副总经理兼"BIM一体化组"总经理　王福魁

一、高科技转化为生产力

2016年是万达商业地产规划设计系统里程碑式的一年。在这一年内,由规划设计系统牵头完成了万达商业地产"BIM总发包管理模式"的研发工作,并顺利进行项目试点。设计中心"BIM一体化组"作为研发工作的牵头部门见证了"BIM总发包管理模式"形成的重要进程。

万达直投项目的规模性开发,其标准化程度高,可复制性强,为BIM技术贯通直投项目"设计、建造、运维"的全生命周期,提供了条件和机遇。万达"BIM总发包管理模式"是在"总包交钥匙模式"基础之上利用BIM技术和信息化手段的创新管理模式,将建筑业项目管理提升至制造业精细化高度,被董事长誉为"全球不动产开发建设史上的一次革命!"。

万达"BIM总发包管理模式"有助于企业的规模化发展,提高工作效率,节省企业成本。"BIM总发包管理模式"的实施可为企业节省管理费数亿元,达成企业"省人、省钱、省时间"的目的,实现高科技向生产力的转化。

二、项目管理模式革命性变化

万达商业地产基于企业多年技术积累,应用科技手段,前后进行了三次项目管理模式的创新。而以BIM技术和信息化为基础的"BIM总发包管理模式",使项目管理模式发生了革命性的变化,具有"管理前置、协调同步、模式统一"的"三大特性"。

产品的标准化实现了"四方"管理和技术工作的前置开展;统一的工作平台实现了"四方"协同工作;统一的管理模式实现了"四方"的"工作标准统一、执行计划统一、操作平台统一、验收成果统一"。

万达"BIM总发包管理模式"实现了项目管控的标准化,实现了真正的总发包管理。

三、设计管理迎来新工作要求

对于设计管理来讲,BIM模型无疑是变化的核心所

I. HI-TECH TRANSFORMED INTO REAL PRODUCTIVE FORCES

The year 2016 is said to be a year of milestone significance for the planning and design system of Wanda Commercial Estate. In this year, the Planning and Design System has taken the initiative to carry out R&D work on "BIM Turnkey Contract Management Model" (the BIM Model) for Wanda Commercial Estate, and successfully done pilot run in projects. Luckily, the "BIM Integration Team" at the Design Center participating in leading the R&D work witnesses the birth of the "BIM Model" all the way.

Thanks to scaled development of Wanda's direct investment projects with high standardization degree and strong replicability, BIM technology has won advantages and opportunities to run through the full life cycle of direct investment projects (i.e. design, construction and operation & maintenance). Proceeding from the "Turnkey Contract Model" and leveraging the BIM technology and information means-based innovative management mode, the "BIM Model" has upgraded the project management in the construction industry to such a refinement degree that is available in the manufacturing industry. This model is thus praised by Chairman Wang Jianlin as "A Revolutionary Change for the World's Real Estate Industry".

The "BIM Model" contributes a lot to propelling scaled progress of a company, improving work efficiency and saving cost in management for hundreds of millions of RMB, attaining the enterprise goal of "saving labor, cost and time" in real sense. Given this, hi-tech has virtually transformed into productive forces.

II. PROJECT MANAGEMENT MODEL UNDERGOES REVOLUTIONARY CHANGES

With years of experiences in technology, Wanda Commercial Estate has technically carried out innovation on the project management mode for three times. Among these innovations, it is the "BIM Model" based on BIM technology and informatization brings about a revolutionary change to the project management mode. The model is featured by three aspects, which are pre-management, coordinated synchronization and unified model.

Product standardization achieves pre-implementation of the "Four-party" management and technical work. A unified work platform achieves the "Four-party" cooperative work. The unified management mode achieves the "Four-party" unification in work standard, execution plan, operation platform and results acceptance.

Wanda's "BIM Model" has implemented standardized project control and the real turnkey contract management.

III. DESIGN MANAGEMENT NEEDS TO ASSUME NEW TASKS

在。BIM模型是"BIM总发包管理模式"的技术核心，利用BIM技术在3D模型基础上输入数字信息，与成本、计划和质量管理信息挂接，满足各部门管理需求。BIM模型包含了30万个构件、10亿条数据信息，涵盖12个分项专业的全专业信息模型，涵盖专业多、覆盖范围广，避免了设计成果错漏碰缺，减少了在项目建设过程中出现的变化，大大降低了项目建造成本。

基于BIM模型，设计管控的重点由原模式下的"全设计周期管控"变化为"变量设计管控"；基于BIM平台，设计管控过程由原来的"单向管理"变化为"四方协同管理"。

针对以上变化，设计管理也迎来了新工作要求——落实变量、控制变更。

1. 落实变量
项目摘牌前，项目公司和设计总包应主要围绕规划变量的复核开展工作，对移交规划条件的准确性复核、项目地方性特殊要求进行调研，为设计开展提供依据。项目摘牌后，项目公司将设计变量调研情况迅速落实为正式的审批意见，避免出现颠覆性意见，对同步开展的设计工作造成重大影响。设计中心对项目变量进行重点复核确认。

BIM模式下，"落实变量"成为规划立项和设计报建阶段的设计重点管控工作，变量的有效落实将大大减少项目建设阶段变更的产生。

2. 控制变更
设计变更必须先发起并审批完成"设计变更申请流程"后方可实施。由变更提出部门发起设计变更申请流程通过后，在BIM平台上同步完成BIM模型、计划节点、成本核算等相关变更确认；变更成果模型落实、平台记录，最终模型与现场交付完全对应。

BIM模式设计变更流程线上审批、设计变更成果线上移交，线下变更手续及文件为无效文件，工程总包有权拒绝接受；如工程总包实施未经审批的变更，将对工程总包进行双倍罚款，相关罚则已纳入总包合同。

BIM模式下，"控制变更"成为项目建设实施阶段的设计重点管控工作。

"BIM总发包管理模式"响应了国家"十三五"的号召——"强化科技创新引领作用"，以科技手段助力企业发展。同时，BIM模式高度契合了国家产业化的发展方向，不断推进标准化以实现产业化目标，承担起应有的社会责任。

For design management, the BIM Model no doubt is the core of the change. BIM models, the technical core of the BIM Model, utilize BIM technologies to input into 3D models the digital information that is related to cost, planning and quality management information and meets the management needs of each department. With 300,000 components and 1 billion data relating to 12 disciplines inside, the full-discipline and widely covered BIM models help to avoid errors, bugs, collisions and defects in design results, lessen variation seen in the process of project construction, and greatly reduce the project construction cost.

Based on BIM models, the design control shifts its focus from the original full design cycle control to variable design control. Based on the BIM platforms, the design control process changes its original one-way management into "Four-party" collaborative management.

Against the above-mentioned changes, design management needs to assume new tasks, including implementing variables and controlling changes.

1. IMPLEMENTING VARIABLES
Before delisting of projects, project companies and main design contractors need to focus on the rechecking of planning variables, and conduct research on accuracy rechecking of the transferred planning conditions as well as special local requirements on projects, which together provide the basis for design work. After delisting of projects, the project companies need to promptly put research results of design variables into formal approval opinions, in consideration of the possible subversive opinions that may lead to substantive changes on the synchronized design work. The Design Center then needs to spare no efforts in rechecking and confirming the variables of projects.

Under the "BIM Model", "implementing variables" grow to be the design control focuses in the phases of planning for project establishment and construction application for design, as its success will slash changes incurred from the project construction phase.

2. CONTROLLING CHANGES
Design changes can be implemented only after launching and approval of the "Design Change Application Process". The implementation process of design changes is as below. First, related changes concerning the BIM models, plan nodes and cost accounting are confirmed on BIM platforms. Then changes results models are implemented and recorded on the platforms. Ultimately, the final models need to be exactly consistent with site delivery.

Under the "BIM Model", the design change process is subject to online approval and design change results are transferred online as well. This means the project main contractors have the right to reject those offline design change formalities and documents which are deemed invalid. In case the main contractors implement unapproved design changes, they will be double fined as set out in the turnkey contract.

Under the "BIM Model", "controlling changes" grow to be the design control focuses in the project construction & implementation phases.

To answer China's "Thirteenth Five-Year Plan" call for cementing the leading role of technology innovation, the "BIM Model" tends to promote the company development with science and technology means. At the same time, the model walks on the development path of national industrialization, in a constant attempt to propel standardization for industrialization and shoulder proper social responsibility.

MANAGEMENT INNOVATION BOOSTS PRODUCT INNOVATION
管理创新力促产品创新

万达商业地产设计中心北区总经理　曾静

自2014年4月份万达商业地产设计中心成立至今，在人员精简而管理职能不断扩大、房地产市场环境不断变化、设计面临越来越大的挑战的情况下，设计中心积极进行管理创新，努力提升管理水平，整合激发各方资源，推动产品不断优化、创新，为营销提供了有力支持。

一、管理创新

1. 管理构架创新
在常规的生产管理部门之外成立总工办及技术部，在人力资源等方面给予相应倾斜；对设计中心进行横向技术管理，在技术层面进行再拔高再创造；各专业总工在产品创新上形成"带头人"，为产品创新提供了管理构架基础。

2. 管理机制创新
形成内部鼓励创新机制，结合考核提升各部门的创新积极性。设计中心内部考核，在创新方案、价值工程、研发专利、案例分享等方面均设有加分项。各部门在正常的设计管控过程中，结合自身工作对产品创新积极思考，不拘泥于传统，百花齐放，百家争鸣。

3. 人力资源及岗位职责设置进行创新
在项目公司层面，制定"区域带头人制度"，选拔优秀突出人才作为区域带头人，给予"带头人"一定"责权利"，促使"区域带头人"形成"专业带头人"作用，在项目公司层面促进产品创新；在设计中心层面，全面实现"全专业项目经理负责制"，打破专业壁垒，激发各专业负责人积极性，打破传统思维框架，为产品创新提供了有力的支持。

4. 管理手段创新
管理利用高科技手段，创新使用BIM管控模式，横向整合各专业及各业务部门工作，提升工作效率及工作标准，为产品创新提供高科技手段。

二、创新成果

通过成体系的管理组织构架及各方面管理创新措施，设计中心产品创新研发成果取得一定成绩，具体如下。

I. MANAGEMENT INNOVATION

In light of expanding management function with personnel downsizing, ever changing Estate Market and growing challenges in design since its outset in April 2014, Wanda Commercial Estate Design Center has strived to be strongly supportive of marketing. It has vigorously implemented management innovation to upgrade management level. It has mobilized and integrated various resources to keep optimizing and innovating products.

1. INNOVATION IN MANAGEMENT STRUCTURE
Besides the conventional Production Management Department, the Design Center has established Chief Engineer Office and Technical Department, for which more human resources are allocated. It has also applied horizontal technical management for technical promotion and innovation. In addition, there is leader of product innovation among chief engineers of each discipline, which constitutes foundation of management structure for product innovation.

2. INNOVATION IN MANAGEMENT MECHANISM
The Design Center adopts an internal innovation incentive mechanism and assessment to encourage innovation enthusiasm of each department. As internal assessment has "huge plus" for innovative solutions, value engineering, R&D patent and case sharing, etc., each department, combined with their own work, is inspired to work on product innovation in the process of ordinary design control work, delivering unconventional and diverse ideas.

3. INNOVATION IN HUMAN RESOURCES AND JOB RESPONSIBILITIES
Project companies of Wanda Group have developed a "Regional Leadership System". These leaders are selected among outstanding talents and granted with certain duty, right and interest to perform the role of discipline leaders. This system thus guarantees product innovation at project company level. Meanwhile, the Design Center has fully implemented the "Project Manager Responsibility System" for Full Specialty. This system helps to stimulate heads of all disciplines by breaking the discipline barrier, and vigorously backs up product innovation by jumping out from the traditional thinking frame.

4. INNOVATION IN MANAGEMENT MEANSL
Adding hi-tech to management, the Design Center ingeniously applies the BIM control model to horizontally integrate work of each discipline and operating department. In the end, work efficiency and job standard are enhanced and product innovation is weaponed with hi-tech means.

II. INNOVATION OUTCOMES

Thanks to the systematic management structure and innovation measures on management, the Design Center has achieved the following R&D achievements in product innovation.

1. 研发工作硕果累累，且全部按计划高质量完成

2016年设计中心级研发共8项、部门级研发10项，专业覆盖面广，研发具有较高深度。研发成果共取得：1项行业规范，出版书籍2本，形成标准模块2项，设计导则2项，标准图集1本，拟申请专利1项。

2. 研发成果落地性强，对项目设计指导意义大

如《南北方植物配置设计标准》研发成果在南北方各项目中均得到了应用，既节约了成本，又确保了效果。又如"住宅底商机电设计标准模块"，在具体项目设计中直接予以套用，既提升了工作效率，又确保了设计质量。

3. 研发紧扣产业化、高科技、社会经济发展趋势

例如《建筑适老化设计研发》、《住宅百变空间研发》、《写字楼万创空间模块研发》，紧扣社会老龄化、高房价区域刚需客群的实际需求，以及"万众创新创业"的小企业写字楼需求，通过研发为下一步产品创新、设计适应市场需求等创造了良好的条件。

设计中心不断推进创新管理，大力推动了各项产品创新及研发工作，有效地支持了营销工作，为完成集团各项营收指标打下了坚实的基础。随着市场的变化及社会科技的发展，我们将继续通过创新管理，创造出更多适合社会需要的好产品，以"设计创新"去创造价值。

1. R&D WORK ACHIEVES FRUITFUL OUTCOMES AND COMPLETED AS SCHEDULED WITH HIGH QUALITY

The year 2016 saw 8 R&D initiatives at the Design Center level and 10 at the department level. These R&D initiatives featuring wide disciplinary coverage and enriched content have harvested the following outcomes, including one industry code, two published books, two standard modules, two design guidelines, one standard atlas and one patent at application stage.

2. R&D OUTCOMES ARE EASILY IMPLEMENTABLE AND OF GREAT GUIDANCE ON ESTATE DESIGN

The Design Standard for Northern and Southern Plant Disposition, one of the R&D outcomes, has been applied in projects of northern and southern China, which saves cost with guaranteed effect. Another example is the Standard Module for M&E Design of Commerce at the Residence Bottom, which can be directly applied in project design to facilitate work and escort design quality.

3. R&D CLOSELY FOLLOWS INDUSTRIALIZATION, HIGH-TECH, SOCIAL AND ECONOMIC DEVELOPMENT TREND

R&D on Elderly-Oriented Design of Buildings, *R&D on Flexible Space of Residence* and *R&D on Co-working Space Module of Office Buildings*, for instance, all closely follow the actual needs of the aging society and the customers with rigid demand in high housing price areas, as well as the needs of office buildings of small business in response to the call of entrepreneurship and innovation by all. These R&D outcomes have created favorable conditions for the following product innovation and market-oriented design.

Proceeding from constantly advancing innovation management, the Design Center greatly impels product innovation and its R&D work and contributes effectively to the marketing work, which paves solid foundation for securing Wanda's revenue indexes. In face of a changing market and developing social technology, the Design Center tends to continue its efforts on innovation management to produce more high-quality products catering for social needs and to create value through design-driven innovation.

图书在版编目（CIP）数据

万达商业规划 2015：销售类物业 / 万达商业地产设计中心主编.
—北京：中国建筑工业出版社，2016.11
ISBN 978-7-112-20181-5

Ⅰ.①万… Ⅱ.①万… Ⅲ.①商业区—城市规划—中国 Ⅳ.① TU984.13

中国版本图书馆 CIP 数据核字 (2016) 第 304593 号

责任编辑：徐晓飞　张　明
执行编辑：李子强
美术编辑：陈　唯
英文翻译：喻蓉霞　王晓卉　郝　婧
责任校对：王宇枢　姜小莲

万达商业规划 2015：销售类物业
万达商业地产设计中心　主编

*

中国建筑工业出版社出版、发行（北京海淀三里河路 9 号）
各地新华书店、建筑书店经销
北京雅昌艺术印刷有限公司制版
北京雅昌艺术印刷有限公司印刷

*

开本：787×1092 毫米　1/8　印张：34　字数：850 千字
2016 年 12 月第一版　2016 年 12 月第一次印刷
定价：1000.00 元
ISBN 978-7-112-20181-5
（29645）

版权所有　翻印必究
如有印装质量问题，可寄本社退换
（邮政编码 100037）